Growing Eco-Communities

Growing Eco-Communities

Practical Ways to Create Sustainability

Jan Martin Bang

Floris Books

First published in 2007 by Floris Books

www.florisbooks.co.uk

British Library CIP Data available

ISBN 978-086315-597-0

Produced in Poland by Polskabook

Contents

Acknowledgments

Ralph Gering estimates there are 350,700 people living in community. Add in all those who have lived in community throughout history and that's a lot of people. I thank you all. Your lives of idealism have given me the raw material for these thoughts and speculations.

The community where I live, Camphill Solborg in Norway, has given me a framework for working on this. Thank you.

My family, Ruth, Jake and Aline, Isak, and Sarah, thanks for putting up with my habit of disappearing in order to do the actual writing. During the final editing of the book Jake and Aline had our first grandchild, Emily, and I dedicate this book, and all my work on communal and environmental living, to her and her generation.

Thanks also to Anne Miller, Tom Bragg and Simen Torp for help along the way. Thanks also to Rolf Jacobsen and Asbjørn Tufto for permission to use photographs. Otherwise all the illustrations in the book are mine.

I owe Christopher Moore a big debt of gratitude for helping me with the editing of the manuscript.

To Christian Maclean, Katy Lockwood-Holmes and Ulrike Fischer at Floris Books, thanks for actually getting the book published.

Foreword

This book is an attempt to give some advice and insights into the life cycle of communities. Looking at life cycles gives sustainability a whole new meaning. Sustainability was originally defined by the Global Ecovillage Network (GEN) as the ability of an ecovillage to continue indefinitely into the future. I now see that most communities don't reach an enduring steady state, just as we human beings don't either. We don't continue indefinitely into the future; we grow, mature, age and eventually die.

I will be looking at communities as organisms, and the hope is that by charting the process of healthy development, this will help groups to avoid some of the traps and pitfalls that beset them on their way.

Because the book will deal with how communities develop, it will have a wider relevance. I have used the terms ecovillage, community, commune, collective, intentional community and village without a great deal of differentiation, but will try to confine myself to the word community throughout this book as much as possible. Each term has subtle differences, but I don't feel the need for exact definitions. What I am interested in is how human groups grow and develop. The book was originally intended for ecovillage activists, but I see now that it's relevant to anyone who works in a group motivated by strong ideals. As I got further into the material, I found a wealth of fascinating insights had already been developed by people working in corporate management theory. I hadn't previously realized how relevant their work is to us who are involved in creating alternative communities.

Intentional community has developed in our Western culture as a means of trying out new social forms, based on the idea that we can create or design social patterns rather than just inheriting them. Western society is what I know, that's where I come from, and that's what I'm trying to improve. The book is personal. It's based on my own biography.

I was born in Norway, and my family moved to London in the mid-1950s where I grew up through the 1960s. I came across the British Commune Movement in 1969, and was involved in various commune groups in Britain throughout the next decade and a half. I moved with my family to Kibbutz Gezer in 1984, and we lived there for sixteen years, with a break for a year to visit communities in England and Norway. In 2000 we moved to Camphill Solborg in Norway where we still live. My working life has progressed from being a schoolteacher to farmer to freelance environmental educator and project manager. In Solborg I work a lot in the administration.

I use Permaculture design thinking a great deal in this book, and it may be as well to give a brief definition for those who have not yet come across this. Permaculture is about designing sustainable human settlements. It is a philosophical and practical approach to land use, integrating microclimate, functional plants, animals, soils, water management and human needs into intricately connected, highly productive systems. It presents an approach to designing environments that have the diversity, stability and resilience of natural ecosystems.

Permaculture looks for the patterns embedded in our natural world as inspirations for designing solutions to the many challenges we are presented with today. Permaculture means thinking carefully about our environment, our use of resources and how we supply our needs. It aims to create systems that will sustain not only for the present, but also for future generations. The idea is one of co-operation with nature and each other, of caring for the earth and its people.

The idea of taking ecological cycles and using them as design patterns may seem obvious when it comes to gardening and farming, a little less so when we build a house, and probably quite foreign to those designing a business, an economic system or a process for group decision-making. It was when Permaculture began to address these problems that it became more relevant, more international, and of greater interest to people involved in environmental design education. This design system did not limit itself to the mechanical and materialistic, but also gave direction to the individual in a personal and positive way.

Friedrich Glasl makes a similar statement about using natural systems as models in his study of business management, *The Enterprise of the Future:*

> If we really want to make progress, we cannot draw conclusions about human beings or the social organism on the basis of an analogy with technology; on the contrary, we have to learn from the biological sciences.

I would like to take this biological metaphor to include the phases of rivers and of plants. This is a classic Permaculture technique, looking for patterns in nature, and one that it shares with more recent developments in Goethean science.

When rivers flow, we can see a pattern in their behaviour. They begin in the mountains as bubbling brooks, leaping down steep stony slopes and over craggy waterfalls. Many geographers call this the youthful stage. As the stream gathers force, and is joined by more streams, it enters a more rolling landscape, it begins to wander from side to side, and to flow more gently. We call this the mature stage. After a while we come to the lowlands, the river can now be huge, occasionally flooding its banks. It meanders to and fro, sometimes leaving curved pieces of standing water known as oxbow lakes. This is called old age, the last stage before the river empties itself into the ocean. Each of these stages, youth, maturity and old age, has its own characteristic landscapes, its own set of plants and animals, even settlement patterns evolved by human beings. It's a powerful pattern and we see it reflected elsewhere: in the life of a human being, the stages of a plant, and in the changes that communities go through in their own life cycles.

It's that pattern of growth, development and decay that I am applying to community in the course of this book. This is not a piece of academic research, and I don't give footnotes or exact references. I do however include a list of the books that I have used as inspiration. It's written for those of you out there who live in community, who might be interested in living communally, and those of you who are interested in how human groups behave. It's based on experience, and I've tried to include as many personal stories as I could. Not all the stories are my own, some I've heard from others.

Two years ago, I finished writing *Ecovillages — A Practical Guide to Sustainable Communities.* That book was based on the Permaculture Design Course, focussing on ecovillage design. Its aim was to help people design and build ecovillages. The following chapters are in some ways a sequel, and broaden the theme to include how small, purposeful communities get going, mature and reach old age.

Now that you've got your community up and running, I hope that this book will help you to nurture it.

Jan Martin Bang
Solborg, Norway

Introduction

Village life is very old. It is a tradition that stretches back into a past so long ago that we measure it by millennia rather than centuries. Richard Critchfield in his monumental work *Villages,* written in the 1970s, traces the first appearance of village life, as human beings emerged from the hunter-gatherer stage in the Nile valley. He comments: 'Villages are man's oldest and most durable social institution, emerging with settled agriculture around ten thousand to fifteen thousand years ago.'

Many people within the intentional community world, me included, would argue that a village sized community is the most natural place for human beings to live. The concept of the ecovillage is based on that very idea and recognizes that we human beings like to have around a dozen close friends, and can make daily contact with up to several hundred individuals before that contact becomes impersonal. Of course there are many today who would prefer the hustle and bustle, the variety and the anonymity of the large city, but these large cities seem to have brought with them some serious social and personal problems. Problems that we tend not to find in villages. However, cities have taken over, and they are here to stay, but Critchfield does observe that at the time he was writing, there were still about two million villages left, predominantly in the Third World.

He makes the interesting observation that 'villages possess a universal culture (based upon tilling land, property, and family), which varies significantly only in the realm of abstract ideas or religion.'

Intentional community and ecovillages do vary significantly because they are based on either abstract ideas or religion, or a mixture of the two. They are qualitatively different from what Critchfield might call normal villages. The difference lies in the intent to set up an alternative to the mainstream society out of which they sprang. This difference is fundamental and deep, being based upon the idea that we can create society, we can design it ourselves, and need not be imprisoned within the social structure we inherit from our forebears. This is the tradition of

Village life has a long tradition

When human beings began cultivating the soil, or finding concentrated sources of food and raw materials, they settled into a village type of life, and formed a tradition that has lasted well over ten thousand years. During this long period, we have evolved social habits which are now entrenched enough to be taken for granted. Throughout the world, villages are now tourist attractions; they are regarded as picturesque and many of us in the Western world dream of retiring into these seemingly timeless, relaxed and harmonious communities.

This quality of tradition has to be balanced against the dreams and aspirations of the individual. A lot of literature about village life has as its theme the rebellion of a single villager against the suffocating traditions of the past. In this dynamic polarization we can experience tension and excitement, tragedy and comedy, defeat and victory.

The positive qualities of such a traditional society are that people know where they stand; there is often security and co-operation built in. In our fast-changing and insecure modern world, these are things we long for, and think we can find them by moving into a traditional village. Certainly, these are the qualities we try to create within eco-communities.

The main difference between these traditional villages and eco-communities lies in our attitudes to change and social creativity. Traditionally, the social form was accepted as being set. Today we question that, and aim to create a new kind of fellowship. This idea, that we can create society out of our free will, is what sets us apart from traditional villages. If we can take the social form of the traditional village, and fill it with the content of creating a better society, we should end up with a winner!

Opposite: Village in Provence

intentional community that has built up in our western civilization over the last couple of millennia.

The village tradition that Critchfield has observed, researched and written about is dominated by continuity and a resistance to change. Intentional community, on the other hand, *is* change: it's about doing something different from what has been. But in our western society today the norm is not the village but the city. Eco-communities are attempts to get back to that more normal, human-sized framework of relationships, at the same time using technology that does less harm to the environment.

In the year 2000, I travelled overland with my family from Turkey to northern Germany, taking five weeks to cover that distance. One of the aims of the journey was to visit ecovillage projects. Our main challenge in planning the journey was to decide which projects to visit, there were so many! Today it's possible to travel the world, and never be far from a community, an ecovillage or a Permaculture project. Eco-communities have penetrated into virtually every nook and cranny — at least in Europe which I know best, but from what I can see, there is a similar story in other continents.

One of the questions which some of us ask occasionally is: how many of us are there out there? How many communities, how many movements and networks? How many people are living like this? Making an assessment of how many people live in community is fraught with problems of definition and contact. In 2002, with his online *Encyclopedia* Ralf Gering made a brave attempt and came up with some exact figures. It was certainly the most comprehensive listing I have ever seen, and his definitions are clear and well thought out. Worldwide he lists 195 organizations or movements, with 3,985 actual communities, comprising 350,700 residents. Not very many, are we, on a planet with six billion people? Amazing, really, that so few can have such an impact.

Looking at how eco-communities have established themselves around the world, I find it useful to differentiate between a network and a movement. Nancy Foy in *The Yin and Yang of Organizations* has given some interesting characterization of the two. She suggests that networks need a focus, a 'spider' at the centre, a newsletter, a list of members, maybe some groups and definitely a phone number, or today better still, an email address.

Movements need clear goals, a chairperson or secretary, a journal, a set of statutes or byelaws, committees and an office building that you can visit. I would add that movements are proactive, with goals that motivate people to gather behind them. Networks, on the other hand, are passive. They are potential openings for action, depending upon the initiative of individuals to use the network resources.

There seem to be two ways in which movements or networks are created. One is by a successful community spawning other communities, which might be regarded as daughter communities of the first one. In my mind I have an image of a truck setting out from the first Camphill in Norway, Vidaråsen, full of furniture and tools, and with a small core group of dedicated people, and arriving at the newly bought farm at Jøssåsen, one of the other villages, and creating a new community. Findhorn members can no doubt tell similar heroic tales, when members decided to start similar projects in other places, like the Sirius community in the USA.

This can create very strong and integrated movements like the kibbutz movements, but there can also be looser connections, more like networks, which I feel is the case with the Findhorn example. It would seem obvious and inevitable that the more integrated and closer a movement is, the more bureaucracy there will be.

In Dan Leon's account, *The Kibbutz,* published in 1964, he described a good example of a tightly knit movement:

> The Kibbutz Artzi is a national movement rather than a federation of autonomous settlements. Through its democratically elected institutions it clarifies and determines general principles and practical policies in every sphere of kibbutz life.
>
> Though it is not a political party, it plays a leading role in the work of the United Workers Party, Mapam, and normally provides about a hundred full-time party workers.

The kibbutz movement was for a long time actually a federation of movements, which has been in a process of amalgamation over the last few years. The biggest movement was called Takam, and comprised about 170 kibbutz communities. Artzi, which Dan wrote about, had about seventy, and the next one down was the religious movement with about seventeen. Each one is now a separate legal entity, leasing the land upon which the communities are built, and determining many detailed questions of policy and finance. There is a staggering amount of movement and federation institutions. Both Artzi and Takam have several serious, well-funded and prestigious institutions of learning, with research archives, comparable to universities, and offering a wide variety of educational courses up to university level. In my time we had two office blocks in Tel Aviv, linked by a pedestrian tunnel, which housed the national administrations of Takam and Artzi, and included a dining room for members. Many kibbutz communities retained places to sleep over in town; there were many full time people working there, and even more part-timers. Wednesday was the generally accepted day for part-timers to

Kibbutz university

Takam was by far the largest of the kibbutz federations, with about one hundred and seventy kibbutz communities. During the writing of this book Takam joined up with the other kibbutz movements to form one federation, with its main headquarters at Ramat Efal. This site was an attempt to set up an urban kibbutz during the early years of the state of Israel which failed, and the place was given over to a study centre, focussing on kibbutz issues. Over the years it has developed into a large campus, with a conference hall, numerous smaller seminar centres, classrooms and accommodation. A large block of buildings was renovated some years ago, and the main kibbutz office in the Tel Aviv city centre moved out here.

Yad Tabenkin was built as a research centre on this campus, and named after one of the great leaders of the early kibbutz movement, Yitzhak Tabenkin. Today it houses an archive, one of the world's most comprehensive on the subject of community in all its forms. There is an amazing collection of community newsletters, constantly updated and properly filed and accessible. There are seminar and conference facilities, administration offices, a publishing company and so on. Both national and international conferences are held here. The central coordinating office for the International Communes Study Association is located here too, as is the International Communes Desk of the kibbutz movement, which maintains contact with communities throughout the world, and publishes the magazine *CALL (Communes At Large Letter)* several times a year. There are connections with universities, both in Israel and abroad and many courses are offered.

If any of you want to study community in all its myriad forms, this is probably the best place to do your research. The picture opposite shows the entrance to Ramat Efal.

come in from various far-flung communities, and there would be an exciting buzz throughout the building.

Saadia Gelb remembers it very differently:

> Every kibbutz has always jealously guarded its individuality and independence. Today, technological and industrial imperatives, common sense, economic pressures and waning local patriotism are altering the scene:
>
> — two kibbutzim have merged their laundries, kitchens and schools.
> — three kibbutzim have combined their irrigation equipment factories.
> — a kibbutz asphalt producer in the north united with another kibbutz producer near Tel Aviv to be closer to the market.
>
> The founders, who dreamed of small, intimate colonies must be amazed as they view us from heaven. The 'Mega-kibbutz' has surfaced.

However one remembers things, the fact remains that the kibbutz movement overall became one of the strongest community movements we have on record. It prefigured the state of Israel by several decades, was originally created in a backwater of the Ottoman Empire, survived the First World War, and grew to strength during the period of the British Mandate despite many restrictions and obstacles being placed in its path. It was highly instrumental in the establishment of Israel, and was one of the major elements in absorbing the hundreds of thousands of survivors from Nazi persecution, and to a lesser degree, the half million or so refugees expelled from Arab countries in the early 1950s.

We can find another good example of a movement in the Hutterites in North America. Joanita Kant gives a very good succinct introduction to the Hutterites, their history and development, in *The Hutterite Community Cookbook*. When you've had enough of reading history, you can turn to the recipes and make lots of great meals! She writes:

> In 1988 there were 35,000 Hutterian Brethren living in 374 colonies or 'Bruderhofs' (places of the Brethren). All but a very few of these are located in the Great Plains area of Canada and the United States. In the Midwest and West there are usually between seventy and 150 people in each colony with an average of about seven people per family. When the population of a colony grows too large to maintain a sufficient amount of work for each member, the colony is divided and a new colony is started.

They are an excellent example of a movement arising from the creation of new communities. They really are like a plant establishing itself and gradually colonizing an area.

How very different from the Communes Network developed in Britain in the late 1960s. From a handful of activists setting up a series of associations, by 1968 there emerged a properly constituted Communes Movement. In that year it had a membership of twenty-four individuals, and their magazine a distribution of forty-six copies. Hardly a mass movement. But these were heady and exciting times, hippies were swinging, students in Paris were setting up a Commune and throwing cobblestones at the police, and there were full-scale confrontations in the USA between anti-war protesters and the authorities. Alternative types flocked to communes. By 1971, membership of the Communes Movement had risen to 340 and there were three thousand copies of the magazine being printed.

The movement was ambitious, its aim was to create a federal society of communities, and a fund was set up to buy property. But the momentum was hard to sustain, and the impulse petered out by 1975. It had become very cumbersome. Voting on all issues was by postal ballot. There was no common ideology, what motivated people to join was the rejection of the existing society rather than the creation of a new one. There arose a split between those committed to the idea of a close-knit federation, and those who just wanted to encourage the setting up of communes and other alternative projects. The second group prevailed, and energy, money and organization capability drained out of the movement, which turned into a network.

Nick Saunders chronicled this process in *Alternative England and Wales,* published in 1975:

> The Communes Network arose out of a meeting in February 1975 to replace the Communes Movement. The Communes Network has no constitution, it only exists as a means of linking people involved in alternative lifestyles.

Ten years later Keith Bailey commented in *The Collective Experience,* a collection of the best articles from both the communes magazines and newsletters:

> The main aim (of Communes Network) was to produce an informal newsletter, a kind of open letter between friends, to allow people to keep in touch and co-operate.

The founding of the Global Ecovillage Network

In 1995 the Findhorn community in the north of Scotland invited people around the world to a conference entitled 'Ecovillages and Sustainable Communities'. Registration far outnumbered possible accommodation, and hundreds had to be turned away. Nearly five hundred people gathered there, using every available lodging for miles around and an enormous marquee as a dining hall.

Most of the people who came were active environmentalists already living in, or associated with, community. This was a classic case of creating a network from the grassroots, rather than imposing something top down. Since there were so many different kinds of community groups represented, there was no chance of imposing uniformity. The definition of what an ecovillage or a sustainable community might be was in the end left pretty vague. The aim of the network was to support and encourage initiatives and projects. This led to a built-in need to embrace variety, something which is echoed in natural ecological systems, where diversity gives an added strength and resilience.

This quality of diversity has given the Global Ecovillage Network a great deal of momentum. Within a decade, these discussions at Findhorn have resulted in a truly global network, an educational system up to university level, and hundreds, if not thousands, of training centres throughout the world.

Ecovillages are here to stay!

OPPOSITE: GEN gathering at Findhorn

The Global Ecovillage Network (GEN) is a good example of an idea linking together lots of already existing communities and giving them that extra boost of strength and motivation that a network can so often do. Jonathan Dawson has recently published a very good book on ecovillages where he writes:

> GEN's main aim was to encourage the evolution of sustainable settlements across the world, through:
>
> — Internal and external communications services;
> — Facilitating the flow and exchange of information about ecovillages and demonstration sites;
> — Networking and project coordination in fields related to sustainable settlements; and
> — Global cooperation and partnerships, especially with the United Nations.

GEN was established in the early 1990s by a number of existing communities and projects getting together and agreeing on a set of common aims, and a loose definition of what an ecovillage might be. This in turn stimulated others to set up new ecovillages, and that in itself is enough to justify the existence of the network. By its very existence, GEN was able to generate funding for member communities, and for activities that brought them together. Within a decade of GEN being founded, a university level training programme has been established, based around Findhorn's educational infrastructure, and recognized and supported by the UN.

Networks, movements, organizations, directories, conferences, gatherings: there are many ways of getting together, and many committed individuals who want ecovillages and communities to work closer together in cooperation. Looking around at the networking of communities and movements, we can find a good example in North America. In the 1991 edition of the *Directory of Intentional Communities* there is a really good collection of articles under the heading 'Networking.' These trace the history of the Fellowship for Intentional Community (FIC) from a beginning in 1940 with the establishment of Community Service, Inc., through a number of related organizations over the following decades. At the time of publication, in 1991, the FIC had over forty member communities and ten associated networks.

Fifteen years later, in the 2005 edition of the directory the picture has changed and expanded. Laird Schaub has given an overview by listing the number of communities in each subsequent edition of the directory:

1990	304 communities
1995	565 "
2000	585 "
2005	614 "

It has to be borne in mind that for many reasons this represents only a fraction of the complete picture. Many communities were not known to the directory editors, many others don't want to be listed. The same directory in 2005 listed seven networks of intentional communities in North America. These directories, published every five years or so, contain a wealth of material about community, in addition to the list of communities, with detailed factual information about each one. As a documentary history of the commune scene in North America, they are invaluable.

Thinking even wider, there have been several attempts to try similar things at an international level, long before GEN emerged in the mid-1990s. In 1979, the Communes Network in Great Britain held a gathering at Laurieston Hall in Scotland, and founded the International Communes Network. This was followed by similar gatherings throughout Europe, seven over the subsequent six years. At the same time International Communes Desk (ICD) of the Kibbutz Artzi Federation organized an international conference in 1981, after a newsletter which had begun in 1976. A subsequent conference in 1985 in Israel launched the International Communal Studies Association (ICSA), which has hosted gatherings every four years ever since, in various venues throughout the world.

Sol Etzioni, the secretary of the International Communes Desk describes the ICD as follows:

> The International Communes Desk is a contact office between the different forms of communal living around the world. The word 'commune' in our

> name is there for brevity, not for exclusivity. It covers communes, kibbutzim, co-housing, ecovillages and other intentional communities. Housed in Yad Tabenkin, we work in close harmony with the ICSA office and, indeed, we complement each other's activities. The members of the Desk come from the various types of kibbutzim and urban communes in Israel. We produce a biannual journal, *CALL (Communes At Large Letter),* which is written for, by and about communities. Since the unfortunate demise of the magazine *Kibbutz Trends,* it is the only regular publication in English which deals with kibbutz topics.

The characteristic of reaching out and establishing contact with others can be identified in Friedrich Glasl's third and fourth phases of the development of organizations, the 'Integrated' and the 'Associative' phases. Here the organization, be it an enterprise or a community, links itself up with other organizations to create a web, many strong relationships to other organizations, stakeholders and environments. Similar patterns can be found in nature, with plants colonizing new areas appearing first as isolated individuals, the seed being transported by serendipitous means. As the number of individual plants increases, and as each individual grows, the result is a network or closely linked mat of vegetation.

Nancy Foy observes a progression of living systems which we might well link into this idea:

> I take ... this belief from another conceptualizing writer, James G. Miller. His concept of living systems works at seven levels: the cell, the organ, the organism (which includes human beings), the group, the organization, the nation, and the super national body.

If we agree that we, people, are the 'organism,' then the community would be the 'group,' the network or organization would be the 'organization,' and the network of organizations might be the 'nation.' I'll leave you to speculate what the 'super national body' might be.

Margarete van den Brink constructs a fascinating and detailed analysis of how groups develop and grow in *Transforming People and Organizations*. She goes on to state categorically:

> All development, be it in an individual, between people, in collaboration situations, in organizations or wherever, takes place according to fixed laws which point the way — a way in which something becomes visible and can be experienced.

She further elaborates a development in which human groups and organizations go through quite definite steps, one leading on from the last:

1. Theocratic organization.
2. Autocratic organization.
3. Bureaucratic organization.
4. Transforming organization.
5. Organization based on moral principles and values.
6. Organization as new community.
7. Organization as contributor to world development.

This is pretty strong stuff. I agree that there are patterns, but personally I fail to find such fixed laws as she does. Looking at communities and ecovillages, I find there are tendencies and impulses, trends and suggestions, but I don't find fixed laws. Her sevenfold development goes into much more detail than I would allow myself, and even Charles Handy's study of organizations with his four stages named after Greek gods (see Chapter 7) tries to keep things flexible. However, I don't want to pour cold water or any other criticism upon Margarete's work; she has done a tremendous service to the study of group development, and opened up valuable avenues for further speculation. And I love her ending; what a positive and hopeful note to describe the last phase of a community as 'a contributor to world development.' You might want to compare her seven periods of group development with the seven-year periods of the human being's life.

But in this book I'm not talking about finding fixed laws. I'm trying to establish new connections, opening up to speculations and wondering if we can find patterns that make sense or that give us insights. For me the wild card, the joker in the pack, is the tendency for communities to change members. Old founders and other stalwarts leave, new hopefuls join, and the group goes through a life change. It's more like the game of snakes and ladders. You get way up there towards the winning space, when whoops, you land on a snake and slide right down to near the beginning again. Communities do exactly that, members often leave in groups, or several key members leave over a short period of time, often for different and seemingly unrelated reasons. New members also often join in groups over short periods of time, again coming from different directions and seemingly unrelated. But the fact remains that over a relatively short period, a number of key members can be replaced by new movers and shakers, and the community takes on a new lease of life. Not all communities experience this, but many do.

In this context, I really like Ed Schein's emphasis upon culture, presented in *The Corporate Culture Survival Guide.* Here he shows how an organization develops its own identity through creating its own culture:

> If the organization continues to be successful and if the founder or founding family is around for a long time, the culture evolves in small increments by continuing to assimilate what works best over the years. General evolution involves diversification, growing complexity, higher levels of differentiation and integration, and creative synthesis into new and higher forms. Specific evolution involves adapting specific parts of the organization to their particular environments, thus creating subcultures that eventually have an impact on the core culture.

The culture that develops in a community is quite specific. It is made up of a myriad of small details, the decorations, the style of building, ways of doing things, the words that evolve to describe both places and things that happen. My perception of this cultural identity is often very subjective, even intuitive. When I revisited the community at Laurieston Hall after a gap of twenty-five years, I felt that it was the same community. When

I visit other Camphill villages throughout the world, I recognize their culture, it feels familiar, I feel at home.

What fascinates me most of all is the tendency for communities to retain their sense of direction, their identity, what Ed Schein calls their culture, despite the changes in personnel. It was exactly this fascination that motivated me to embark on writing this book. What is it that lives in community that survives the passage of time and the changes in membership?

With all honour and respect for Charles Handy's four Greek gods, and for Margarete van den Brink's sevenfold division, I would like to present my own simple threesome: Youth, Maturity and Old Age. I offer this not as a scientific analysis of the development of community, but rather as a starting point for speculation and discovery. Keeping in mind the image of the plant and of the river, which also go through developmental stages, what I want to do is to try to make comparisons between different objects in their growth. Can we see similarities? Can we see differences? Can we learn something from this which might help us while we travel through our own development in our own community? My aim in this book is to raise questions rather than to answer them, to get you to think about the community that you live in, and how it functions. Ultimately, how you can contribute to its growth and development.

Part 1

Youth: the Pioneering Phase

Intentional communities and eco-communities show a number of characteristics in their early phases. The purpose of this first part is to illustrate a number of these from real life examples. This will cover aspects of work, decision-making, finances, membership issues, building and farming. This is setting the stage, this is where groups start out. The way in which they tackle some of these initial challenges will determine how they develop into maturity.

The pioneering era is often the most exciting period of a community's existence, and memories of it will linger on with nostalgia. So much so that there is often a desire to get back to the good old days. But these are gone, and though the wish to remain young may well be genuine, communities, like a human being, also grow and mature, and the behaviour that was fine when young, becomes inappropriate in someone older.

Friedrich Glasl in *The Enterprise of the Future* identifies four phases through which organizations or communities might pass. The first phase he calls 'the Pioneer Phase.' In this first part we will look at some of the characteristics of this phase, and follow it up in subsequent parts to see if we can find patterns of development.

By following a central idea in Permaculture and tracing patterns from the natural world onto things that we are designing, we might draw a parallel between communities and compost. When we want to start or nourish a garden, we make compost and apply it to the soil. Compost is food for the soil, ultimately giving us healthier plants and higher yields. The composting process can be of two kinds, aerobic, an air filled process which results in healthy, nourishing fertilizer, and anaerobic, where the air is excluded, which leads to a smelly, rotten substance with very little nourishment. Allowing air to circulate freely amongst the raw compost is like allowing ideas to circulate freely amongst the core group of a new intentional community. To be closed to new ideas, to exclude the world around us, may well result in a sticky mess of a community.

Another characteristic of the composting process is the chaotic state of the mixture in its initial stages. Compost is best made by mixing up a number of ingredients, stirring them up well from time to time. In this chaotic state, large numbers of micro-organisms multiply rapidly, which in turn create humus out of the raw biomass we started out with. The initial stages of a community might also be experienced as chaotic, ideas come and go, things happen fast, opportunities present themselves and need to be utilized. Hectic days, long hours. Stressful but exciting.

At the same time we need to keep our heads, to think far into the future of possible scenarios. It's important to keep options open and not get locked into things. The importance of laying good foundations from the beginning cannot be emphasized enough; these will ensure longevity for the community. One of the core aspirations which characterize communities is the aim of creating a better society, social altruism, if you will.

1. Everybody Does Everything

In the early years of Kibbutz Gezer, there was a large board at the entrance to the dining hall. On it were horizontal tracks, each one denoting a branch or workplace. Every member had a wooden counter with his or her name on it, and every evening the work organizer would shuffle the counters around according to the needs of the different branches. Next morning, when you came in to breakfast, you could check where you were working that day.

A fairly simple system, one that gave a lot of flexibility, and typical of a young, idealistic community. Most people knew how to tackle most jobs. As each branch was started up, many people helped, and got to know what to do. Training was given on the job, and routine jobs were covered by a larger group of people trained up for it. In our milking list we had 'A' milkers, who really knew what they were doing, and 'B' milkers who needed an 'A' to work with, but had enough training to be of real use. There were experts and specialists, even at this early stage, but they were able to take a day off their specialities to step into other roles when the need arose.

This is a typical situation among young idealistic collectives, communities and co-operatives, and ecovillages are no different. When the ideal is paramount, and the motivation to make money is not to create personal profits but to sustain the group, people are prepared to do what needs doing. This characteristic is even more pronounced in a community which has radical income sharing. When there is no monetary value attached to work, it really doesn't matter where you work. All tasks are performed for the community, not for one's personal gain, and everyone lends a hand, putting their shoulder to the wheel.

Walter and Dorothy Schwarz, travelling round in the 1970s, found a classic example of this at Crabapple, one of the small rural communes set up in Britain in the early 1970s. In their book *Breaking Through,* they write:

> After initially sharing out jobs on roster, Crabapple moved to increased specialization. Some people run the farm, others run the shop, yet others specialize

> in building rather than farming. However, everyone works once a week in the shop. One member takes charge of cooking. Two of the members draw the dole, on the grounds that they are not members of the farm partnership, some have outside jobs, 'to get a bit of air' and these earnings are pooled; everyone gets £5 a week pocket money.

This is a classic characteristic of communities in their start-up phase: work is shared out fairly informally, and everyone lends a hand where it is needed.

A quarter of a century later I found myself in close contact with Hurdal ecovillage in Norway, founded in 2002. The people who set up Hurdal wanted each family to have as much freedom as possible, and be responsible for its own economy. The idea of income sharing was not part of their vision. Even here, though, the task of building the first houses became a shared job, people lending each other a hand on an informal basis. Most of the first houses were built with strawbales, a technique where a great deal of the work can be carried out by unskilled labour, as long as there is good motivation and co-operation. A lot of building skills were learnt and shared during these first years.

When Ruth and I were involved in starting the Louth Wholefood Co-operative in 1980 in Lincolnshire in England, we were a much smaller group, and we never pretended to want to live together. We were motivated by the idea of making wholesome food available to a wider public, and to the ideals of co-operation, of sharing ownership, group decision-making and equality. Everyone took their turn behind the counter, ordering was co-ordinated by Ruth, but most of us knew how to help with it. Derek was the best carpenter, and cranked out shelves and counters almost overnight. John and I found ourselves having to learn accountancy and book-keeping, not because we were particularly good at that, or even enjoyed it, but someone had to do the dirty work! The main work, of standing behind the counter and selling to the customers was done by every member, we could all open up the till in the morning, and tot up the takings at the end of the day. We all kept the shelves stocked, we all kept the place tidy and we all swept up.

Information centre at Kibbutz Gezer

The entrance lobby to the dining hall was the place where we members had the most interaction. We passed through at least six times a day, on our way to and from meals. It was there we ran into others, chatted, and read information from the numerous notice boards. Each member had his or her own postbox, where of course the post was delivered, and where we could keep all kinds of little diddy personal things. Worklists were kept there, and the work board that had each member's name tag, moved around by the work organizer as needs changed in the different branches. We shared cars, and there were also the car lists, telling us who had the different cars, when, and where they were going.

For me, the notice board of any community is one of the most fascinating things to check out. It tells so much about the fellowship, how they organize themselves, what are current issues, what people do. When I gave tours of the kibbutz, something I did a lot when I worked for our seminar centre, I always included a stop at the car board to explain how we shared our cars. Most people were impressed: how ecological, how highly social, and so on. When a visitor came from another community, with decades of communal living behind him, he took one look at the car board and commented: 'Well, in our community we do it very differently, but the main aim is the same. This board tells you who to be angry with when the car doesn't get back on time, right?'

Kibbutz Gezer postboxes

Building the ecovillage at Hurdal

When the first group moved to Hurdal ecovillage, there was one house on the site, a small family house with about three bedrooms. Some people brought trailers to live in, others crowded into that one house, which also served as a kind of communal dining room, kitchen, meeting room, kindergarten, public toilets and shower. Talk about crowded! Building work started as soon as possible, and a variety of different building techniques was employed from the start. Strawbale housing, a log cabin, a prefabricated wooden house, and two experimental wooden houses were soon all under construction. People helped each other in all kinds of ways. Clearly it was in everyone's interest that houses were built as quickly as possible.

This practice of sharing the work was carried on by visitors also. We had a Scandinavian Permaculture gathering there in 2003, and participants were drafted in to help do some of the building work. There were lots of work gatherings on site, courses were held in strawbale building techniques, and building skills were acquired.

Hurdal ecovillage, house building

During the first year there was hardly any money left over for wages, but the shop opened on time, and the turnover kept increasing. Today, twenty-five years later, the shop is still selling wholesome foods, it's still a co-operative, it now owns its own premises, and members have been earning a living by working there for many years.

One of the characteristics of new groups working together is that most of the members can tackle most of the jobs. This is true of ecovillages, communities, collectives and many close knit groups and organizations at an early stage of their formation. As time goes by, this characteristic changes, it evolves, the community or group matures into specialization.

In some cases the practice of such shared work becomes enshrined in the principle of rotation. Here, specialized tasks are held by individuals for limited periods, and then turned over to someone else. This has the advantage of a built-in learning curve, helping individuals to appreciate different tasks, and not to be impatient when others find it difficult to carry out certain tasks. It also means that really tough or unpopular jobs get done, even when no-one wants to do them. This is really important when there are no wages or salaries paid for work done. You can always make an unpleasant job more palatable by paying people really a lot of money for doing it! The job of work organizer on Kibbutz Gezer was just such a job, and even years later, when the community had stabilized and the founders had mostly moved away, this job carried a six-month duration, and it was acknowledged as a tough cookie that everyone sooner or later had to deal with.

On Twin Oaks, in Virginia, USA, they practised a job allocation system whereby people could do what were considered unpleasant jobs and get extra time off for those jobs. In a way, they were paid by getting more free time instead of more money.

The disadvantage with rotation, the reason why it cannot be practised in every job, and why it fades away in many groups, is that it is highly inefficient. We'll see later why specialization gradually becomes a stronger feature of community life, but it is built into the nature of things that as we gradually learn the skills necessary for a job, we get better at it. If after six months I have pretty well mastered the problems of how to organize the work distribution and I am replaced by someone new, it means that

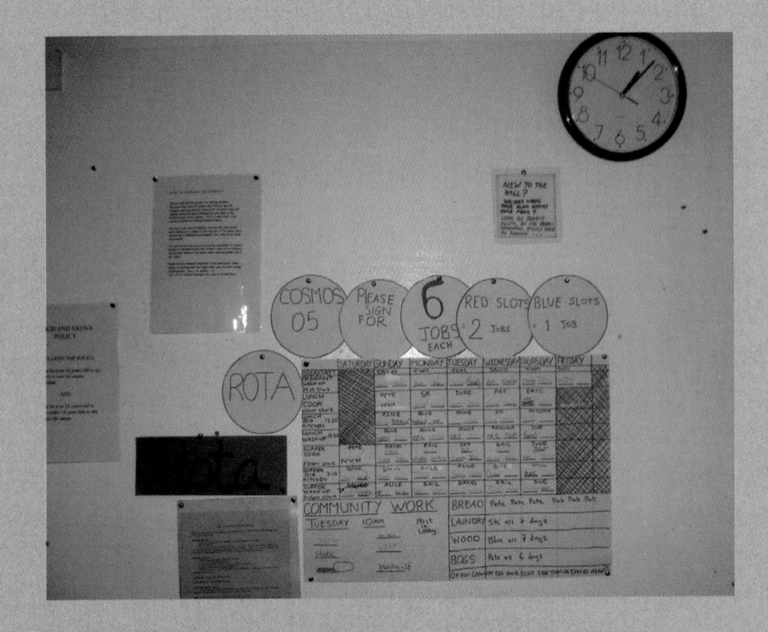

Work schedule at Laurieston

Work schedule culture

Every community will work out its own way of organising how work is allocated internally. A mixture of pragmatism and ideology, the work schedule will fairly quickly become part of the everyday culture of the place. As a commune member, the work schedule and allocation will become a central part of your life, something that will take up a significant part of your conversations with other members. An endless source of fascination and frustration. You never quite get it right! There are always gaps and holes, things get forgotten, people forget to inform about their days off or holidays. Some get overworked, and others will be suspected of slacking.

I always find it fun to look at work rotas when I visit communities, and I always wonder what it all means. Each place has worked out its own special language. I snapped this work schedule at Laurieston Hall in 2005. They've been going for over thirty years, and have worked out their scheduling through many changes. I did wonder what Cosmos 05 meant, and what it really means when you sign up for red or blue slots. This is deep insider stuff!

work organization is always being done by someone who is learning. Very often we learn by our mistakes, and so the organization of work is continually a big mistake. Clearly inefficient from a mechanical point of view! But it has tremendous human potential in the individual learning about themselves and others.

The principle of rotation can in some cases be taken to the extreme. In the early 1970s I used to visit a community in East Anglia in England. I never lived there, but a close friend did, and I enjoyed doing odd building jobs on the old and decrepit farmhouse the group had acquired. It was there that I learnt to mix cement and do rendering!

A conversation that developed over dinner one evening was about care of the goats. My friend, we'll call him John, had more or less taken over caring for them, and was accused of making them 'his' project. There had been no formal agreement over division of labour, but it was important to some members that no-one 'owned' any of the property or the activities. Anyone seen to be too busy in one area was suspected of taking over that activity and creating a feeling of ownership. John was accused of bourgeois revisionism, Proudhon (property is theft) was quoted, and he barely escaped serious re-education. These were early, pioneering times, not just hard work physically, but learning to operate as an idealistic group was really breaking new ground. There were few guidebooks around. Had we then been aware of how this characteristic of everyone sharing work was really a common phenomenon of an early stage of community life, we might have been more relaxed about it.

The Camphill Network of ecological villages was created in the 1940s by a group of Austrian refugees from Nazism who were inspired by the Spiritual Science of Anthroposophy developed by Rudolf Steiner. The Camphill Schools began by working with children with mental handicaps in 1939, and subsequently developed to create communities where adults with mental handicaps worked alongside other people, co-workers, to support themselves. Throughout the world today there are more than one hundred Camphill communities in over twenty countries. A similar process of everyone sharing the work that was to be done developed amongst the pioneers here too. Friedwart Bock wrote about the early years of Camphill:

> Once they had found their way to Scotland the eight young men and women around Dr König had to learn how to tackle practical tasks which they had never before faced in the shelter of their previous homes. They had to learn to serve the needs of the growing community.
>
> Thomas Weihs, a medical doctor, took on the farming work together with a group of delinquent boys. Anke Weihs-Nederhoed, a dancer, took on the laundry with great energy and dedication. Peter Roth, a medical student, taught the children; and Alex Baum, a chemistry student, learnt the rudiments of gardening so that the community could live off the produce of the land. Carlo, an artist, acted enthusiastically as cook.
>
> Alix Roth, a trained photographer, took on the study of nursing. Hans Schauder, a medical doctor, gave music lessons and led the choir. Tilla König and Barbara Lipsker were model mothers to their own families and became the home-makers of the community for the children and the co-workers.
>
> In 1940, Karl König wrote about the truly wise person who knows no specialization and will do any task asked by the community. He or she will endeavour to bring order into life.

You can read more about these fascinating pioneers in the book *Builders of Camphill.*

On Kibbutz Gezer it was for many years the tradition that new prospective members would work their way round a number of branches in their first year, before settling down in a particular enterprise. It was close to a principle that you did not carry on working in whatever profession you had before you arrived. I spent the whole of my second year on Gezer working in the kitchen, even though my professional background when I arrived was in agriculture. This relationship to the kitchen persisted all through my kibbutz years. The day before the Gulf war broke out in 1991, all the American volunteers were flown out of Israel on the orders of the US embassy at just a couple of hours' notice.

I had spent the morning, from 4 o'clock, making cow food out on the farm. When I got in to breakfast, I was asked to step into the kitchen to make pizza for over two hundred people, because half of the kitchen crew had disappeared overnight. It was no problem; everyone got pizza for lunch. My kitchen background proved itself, and everyone was fed.

Years later we had set up a seminar centre where I then worked, and we got a request to make breakfast for five hundred American students. One of the founders of the seminar centre, David, also had a kitchen background, and again, it was no problem, we got up at around two in the morning with a full kitchen crew, and breakfast for five hundred was served at nine. The logistics of this particular exercise was so complex that we had six people just directing buses when the visitors arrived! Again, our experience in different branches helped save the day.

When people can step into jobs at short notice, it gives tremendous flexibility in the working life, and ensures that jobs really get done, even when other tasks intrude into the time of some individuals. This can be very valuable at any time during a community's life, especially at the early stage, when some individuals need to be freed to have meetings with planners and other outside bodies, but don't have enough work to give them a full time job in administration.

There may be a balance here between short term and long term efficiency. Had I gone straight into farming, and stayed there for many years without experiencing other work branches, I would probably have been of great value to the farm. As it was, I gained other experiences which in time proved to have a value of their own.

But why is efficiency so important? Is that the reason we move to pioneer ideological communities?

When I first spent any time in Camphill, in Vidaråsen in Norway in the early 1990s, I was asked to organize the moving of a large store of candle making equipment from one place to another. I knew that it would have to be moved again soon, and felt annoyed by this lack of planning and efficiency. I complained bitterly to one of the long term co-workers there about how inefficient the whole thing was, about lack of planning, of long term thinking, of orderliness and strategy. She listened patiently to my whining and moaning, was silent for a moment, and then asked me: 'Why do you think we are here? Why do we live together in community? Is it to be efficient and orderly?'

I can't remember my reply, but I'm sure it wasn't very coherent, philosophical or spiritual. The conversation had its effect, though, and it is worth pondering how important efficiency really is, even if you're from northern Europe.

Flexibility and the ability to tackle complex situations have tremendous value for the new community. Shared experiences, the success of handling complicated challenges, and the feeling that everyone is pulling together, these are some of the things that create that special emotion of building a group feeling. This is spiritual nourishment for the group. Rotation of work is often thought of as a way of practising direct democracy, and of ensuring that people's egos don't carve out comfortable niches and protected little enclaves. That may be true, but I feel that the real value lies in confronting the inherent disorderliness of life. It gives us an opportunity to create order out of chaos.

At the beginning of a community's existence, there arises a situation which can be compared to the vegetable garden. Mulch and compost is applied to the soil, everything is mixed up and things begin to happen. The micro-organisms burst into a varied and explosive anarchic mess. Compost heaps heat up to such a degree that it has been known for them to burst into flames, just like I did when confronted by having to move candle equipment. This is healthy, the garden soil needs that messy composting activity to create nourishment and eventually stabilize into life giving humus. We need it too, in our communities. Shared work builds community, and the memory of it is often embellished to create the 'golden age' legends which in later life sustain the community.

This creates the community glue, the feeling that everyone is working together, and is of tremendous value both at the time, and in later years. A community has its own biography, its own life story, and what happens at the beginning, during those formative months and years, will colour its subsequent growth and development. Each individual member will create direct, personal links with many different areas, giving them a feeling

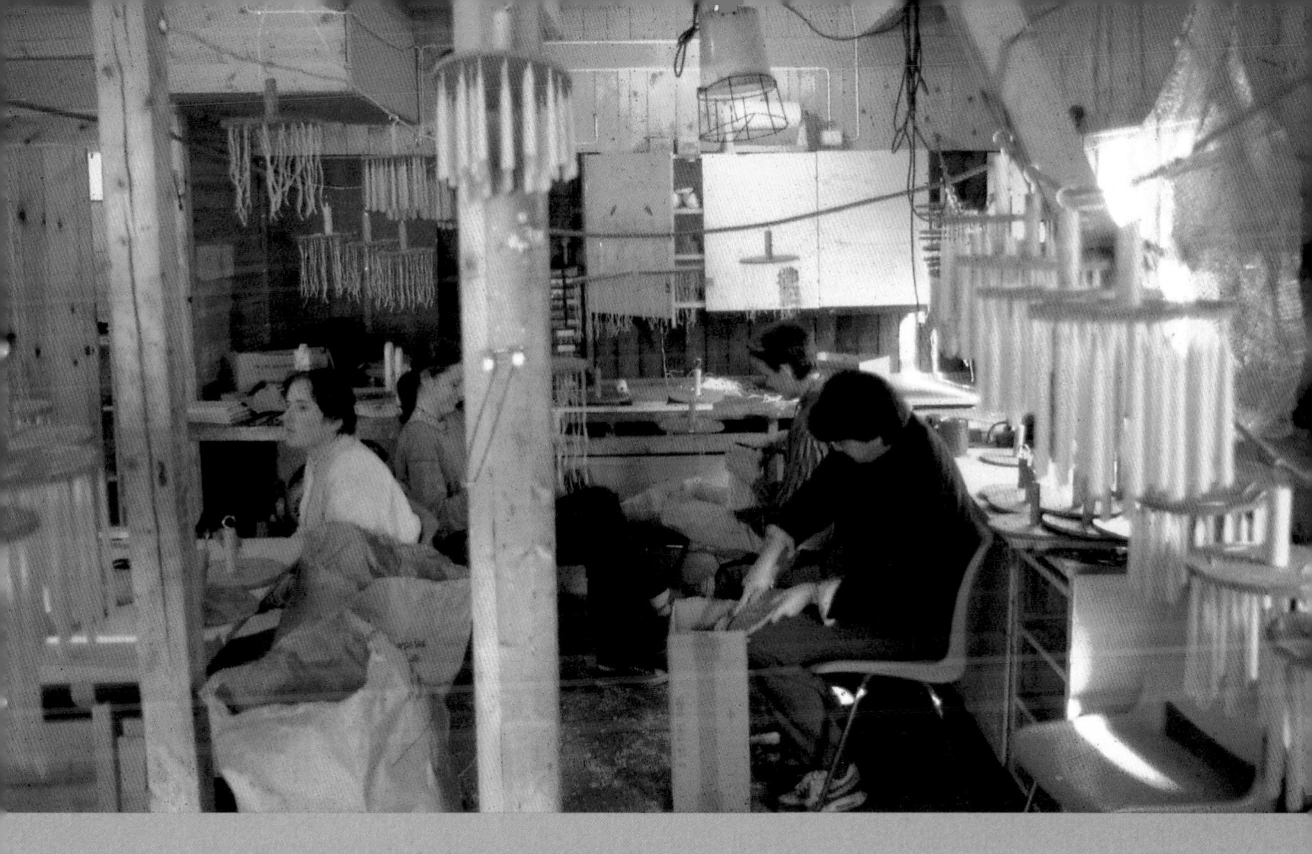

Candle making at Vidaråsen

We lived at Vidaråsen for some nine months in 1991 to 1992. Coming up towards Christmas, it was decided we needed to make candles for sale. Somehow I became involved in this, and helped to set up the candle making equipment, which had been developed there over many years. We soon got going, and the pressure gradually increased as Christmas drew closer. At the high point we had over twenty people working there, and were producing several thousand hand-dipped candles per day!

Candle making was a tradition that began right at the beginning of the community, when they were wondering how to finance the building up of the village. They had heard that high school pupils in Oslo used to collect money for good causes, and had a meeting with the organizers, who were very enthusiastic about helping. As they sat around a table talking about how to raise money, one person pointed at the lit candle in the middle of the table, and said 'How about selling candles?' That was the birth of the 'Light a candle for Vidaråsen' money collection campaign which carried on for seventeen years, and each year enough money was raised to build another house.

that this is ‘my place.’ Stories emerge as part of the oral history which get told as the years go by, to quote Shakespeare, ‘remembered with advantage.’ These stories gradually get embellished and often exaggerated, but nevertheless, gradually become part of the remembered history.

This is how a group builds its own self-consciousness, it begins to write its own biography. Every community is a unique creation by a group of people and just like every individual person, it has its own story. This book will try to find patterns in those stories.

2. Decisions over Dinner

You don't have to live in a community to experience informal decision-making at first hand. We all do it if we live and share our lives with other people. Any family will talk over what they are planning to do as they sit around the table eating together. From the most mundane:

'When are we going to see that film?'

To the more serious:

'Are you really going to make a career in architecture?'

Eating together is an ancient way of getting to know one another. We've been doing it ever since we invented eating and being with others. We probably began before we were *Homo sapiens,* and over the millennia it's become our way of getting social nourishment. The Jewish tradition institutionalized this by creating the Passover meal *(Seder),* a ritual which has been in practice for over three thousand years. Jesus followed this tradition and led the Seder for his disciples before he was crucified. It came to be called the Last Supper by Christians and forms the basis for many Communion services. Eating together is far more than just stuffing food into our mouths.

Many groups, before they ever begin living together, share potlatch dinners, with everyone contributing by coming with a dish to be shared, and during, before, or after the actual eating, a meeting might be held to discuss progress with the project. This often carries on in some way, even if the group, once established, don't eat all their meals together. Co-housing groups for example, often eat together one evening a week.

In the late 1960s and early 1970s there was a vigorous development of communal living in Great Britain. New groups established themselves, enthusiasm was running high. These communities were generally filled with young radicals, and there was a strong feeling of breaking with past social sins, of creating an alternative society. Anarchism and libertarianism were key concepts, and for many of the young people involved, established society was heading for destruction and didn't have a lot to teach. But it was hard to learn new ways; community literature was either stuffy and academic, or consisted of amateur cyclostyled newssheets distributed through personal networks. Clem Gorman wrote in 1975:

> One of the most interesting community projects is Cracker Co-operative, a mainly students' commune established in the East End of London with the aim of stimulating local community action. Once they had moved in, their community life was easy and closely shared. They ate together, always made decisions together, and, though they had their own rooms, spent a lot of time together in the main lounge.

You can read more about this fascinating time in Clem Gorman's account published as *People Together.* Many of these communes were small, intimate groups, which looked more like family than political collectives. There seems to be a pattern here, that small groups resemble family structures in their early years. We find the same insight in Friedrich Glasl's study of business organizations and how they are managed. He comments in *The Enterprise of the Future:*

> ... in this first phase the whole development of management and organizational structures is modelled on the family. The pioneering enterprise is person-centred in every respect.

It's not often that we lump together communities and business management, but at the level of group behaviour, how decisions are made, how projects are carried through, the two have a great deal to learn from each other.

Even with larger groups, the tendency was for smaller subgroups to establish themselves. My own experience of kibbutz life was that meal times were very important for getting things sorted out prior to actual decision-making meetings. We all ate together in the dining room, on Kibbutz Gezer we were over a hundred people. On some kibbutzim there were up to two thousand people eating together. Whatever the size, the natural thing was to relate to the people at your table. Breakfast and lunch tended to be with your workmates; we started work before breakfast, and usually went down to eat as a group. Around the table it was natural to chat about work things, and the end of the meal was often a quick round of who does what for the rest of the day. Quick, easy and relaxed decision-making.

On the other hand, supper on the kibbutz tended to be eaten with your personal friends, but not so much with your family. Once the kids were old enough, they wanted to eat together as a group, not be stuck with boring old grown ups. One of the regular topics of talk around the table were the issues of the day, what kind of choices were facing the community, and what our response was to be. Here issues were thrashed out, opinions were exchanged and weighed up, and individuals often began to form their own responses.

In later years the dining room closed and families were forced to eat at home separately. Sometimes smaller groups would get together to eat, these were often splinter groups who gathered together round an idea or an issue, and there was not the same opportunity to meet people you disagreed with. Whether this was a contributory factor to the conflicts that developed at that time is hard to say objectively, but my feelings at the time, and these are reinforced by later reflection, were that there was a loss of unity. One of the comments that I hear from my friends still living there is that the sense of community has been lost since the dining room closed.

The natural tendency to chat informally and sort things out over a cup of coffee will always be there, and should be encouraged rather than seen as something anarchic and dangerous. However, decisions that are going to affect the future need to be clearly formulated and written down. It's easy in the beginning to be relaxed, but if differences and disagreements arise later, it is useful to have clear statements to refer back to. The best way to combine these two tendencies might be to thrash things out informally during lots of coffee breaks and meals together, and then hold formal meetings when the decisions can easily be made and formulated on paper.

Eating together, kibbutz style

You really get to know people when you eat together three times a day. In the early days on Kibbutz Gezer, there were only very basic cooking facilities in our houses, and anyway, meals were provided in the dining hall. Everyone took their turn at washing up, so there was very little incentive to eat at home. I enjoyed a certain feeling of ideological asceticism, eating off trays with plastic crockery, rubbing shoulders with people in their work clothes from the different branches. The dining hall was crowded; there were more members joining every year, people kept having babies, and parking prams became an issue. When my mother came to visit, she called it the canteen, and I don't think she really liked eating there that much.

Supper nearly always took a couple of hours. A lot of time was spent talking to people, sometimes moving from table to table in order to get together with different people as they came to eat. And afterwards there was always the crowd outside, sitting around smoking and watching our kids play together. There was no saving in time just because the meal was prepared by someone else, and the washing up taken care of by a rota.

The dining hall really provided the framework for an intense community building activity, a space for social interaction.

In *Creating a Life Together,* Diana Leafe Christian strongly urges groups to define a vision, and create strong decision-making structures at an early stage. This is indeed good advice. She has edited *Communities Magazine* in the USA for many years, and has observed the phenomenon that many groups collapse in their early stages. Those that get beyond the pioneering stage very often have their vision and structure well established early on.

Certainly when it comes to presenting the project to outside bodies, it is vital to have some serious documentation. Securing loans, applying for permission to build, and attracting new members, all require that a group has some documentation to present. Banks lending money want security, local authorities granting building permits need to be persuaded, and prospective investors, either voting with their feet and joining the group, or supporting from the sidelines, need to know what they're getting into.

Every community or group should work out its own way of doing this, one of the important features of a developing group is that it creates its own identity or style. However, we may not want to reinvent the wheel every time, and it is always worth looking at the experience of other groups. Here is a short piece of advice, based on observing what happens to groups in their early stages.

1. It would be good to begin working with the vision, the basic idea that brings the group together and infuses it with life. A vision statement can be short, inspiring and to the point, it can bring the group together and be attractive to prospective new members. It should be short, and as universal as possible.
2. Once a unifying aim has been stated, it would be good to give some indication of how this is to be achieved, but again, a short series of succinct strategies are fine to formulate at this stage.
3. Detailed instructions could be the third and last stage, and this could be the focus of a more detailed and painstaking group work. Often this might be called the 'development proposal' and be the kind of thing which forms the basis of planning or loan applications.

This approach, from the whole to the parts, is one that will be repeated at various intervals throughout this book. It is part of a universal truth that it is easier for most people to think big first, followed by filling in the details later. As in most universal truths, there are exceptions, and in the end, it's up to each group to develop its own style.

Sometimes there emerges a very tight and clear organization in a group, which some bright spark will commit to paper, and which looks very impressive. However, this may be an illusion. In the late 1970s Stephen Gaskin from The Farm in Tennessee visited Great Britain, and inspired a group of people to begin organizing a larger collective, somewhat modelled on The Farm. Several meetings were held, and a vision emerged and was written down. There were four main principles:

— Common ownership of land and houses.
— Income sharing.
— Non-discrimination.
— Decision-making by consensus.

This was not an easy process, to quote from the fourth newsletter of the group:

> ... the meeting started with trying yet again to define our fundamental common aim. As usual, this proved fruitless.

In the summer of 1980 a giant wheel was rolled a thousand miles around England. The group adopted the name Cartwheel, many people became involved, and a great deal of publicity, most of it very positive, was generated. A year later a co-ordinating office had been established, the group divided up into specialist working groups, each with a co-ordinator, and membership was divided up into regional groups. On paper it looked very impressive.

However, it was mostly a façade, existing only on paper, in the newsletters, and in people's heads. Things didn't get done, specialist and regional groups never actually met or kept in touch with each other. The Negotiating Group went to see a site at the invitation of the Sheffield City Council, but the plans they asked for were never drawn up by the Planning Group. When a proposal for organizing the farming was presented by

Many hands around the table

Sharing a meal together is an archetypal social gathering, an activity which stretches right back to the beginnings of human life. There are many ways of making a meal even more special. Grace before and after meals sanctifies the repast; some groups hold hands, others always light a candle on the table. In our house in Camphill Solborg we do both of these. The Bruderhof community sings before a meal; some monasteries have silent meals with someone reading aloud from a religious work.

Mealtimes can also raise social awareness. I have experienced meals where no-one is allowed to ask for anything, the exercise consisting of being aware of others' needs, and offering them what they might need. This can also be done in silence. A bit strenuous if it's done too often, it's actually a great way to be reminded of other people's needs and our own responsibility in fulfilling them. In Camphill we try to set aside Saturday evening as a special meal, preceded by a meditation, and during the meal that follows, we keep the conversation level a bit higher than normal, looking back over the highlights of the past week. We usually set the table with a cloth, and put on our best clothes. The meal becomes a gathering where the members of the household feel that something special is being celebrated. Some call this Bible Evening, as we usually end the meal by reading a short text from the New Testament. Others give it different names: Community Meal, Week Ending.

Community can be created around the table.

the Farming Group, it was thrown out by new members who wanted to keep everything flexible and not be tied down by the written word.

Within a couple of years Cartwheel ceased to exist, but the experience remains as a cautionary tale, one which contains several important lessons (you can read a more detailed account of Cartwheel's short history in *Diggers and Dreamers* from 1989). For this chapter, about how communities organize themselves, it can be seen that even though there might exist an impressive organization on paper, unless each individual involved stays on track, and fulfils their commitment, it can still fall apart. Not such a bad idea to get together over dinner more often, to keep the formal structure going by chewing it over informally.

There are different styles of decision-making. At one extreme we find the community that is dominated by one person, a leader or guru, who conveniently makes all the decisions, and may even own the property where the community lives. Of course this frees everyone else up from having to think and make decisions, and as long as everyone is happy, and the group does not oppress or exploit anyone, including themselves, I guess it's OK. In the process of individualization which is the hallmark of our time, fewer people are satisfied with such a system, and indeed, it often seems to result in unfortunate developments, corruption, community collapse and sometimes violent tragedies. Some form of participatory decision-making seems preferable.

Charles Handy studies and teaches organization development and in his *Gods of Management* he comments:

> An examination of the variety of co-operative organizations operating in the UK in the 1970s reveals that the successful ones are always led by some kind of charismatic, energizing figure. His power seldom stems from ownership, but from personality, ideas and initiatives. He is continually re-authenticating himself, depending on his colleagues for consent.

During the 1970s I felt that one of the tasks of a group was to temper the domination of such individuals by using democracy. Growing up in the 1960s, I inherited a tradition of democracy as one of the inspiring ideals of social process. This was reinforced by working in the Trade Union movement, and by being active in the British Labour party in the early 1980s. During our years on kibbutz, the idea of democracy was central to the decision-making process, but here, during the 1980s and 1990s a new element was creeping in, the idea of consensus. In most working committees on Gezer, decisions were arrived at not by voting, but by discussions gradually finding their way to a decision which was acceptable to everyone present. It began to emerge that this was preferable, and sometimes, though not always (notable exceptions at times!), voting in the general meeting of all members was a mere formality, the decision having been prepared by careful committee work and a great deal of informal discussion.

In the Camphill community of Solborg where I live today, we aspire to consensus decision-making, but do have a fallback position of being able to vote if the need arises. It has often been said that one of the disadvantages of consensus is that it can block decision-making and effectively paralyse a community. The deeper you become involved in consensus, the clearer it becomes that this is not just a word to use, but a group process which needs to be learnt.

Betty Didcoct, writing in Hildur Jackson's book *Creating Harmony,* gives a good, succinct definition of what consensus is and how it works:

> Consensus is about people making decisions together. It is based on cooperation rather than competition and seeks solutions where everyone benefits. Consensus decisions build unity from diverse viewpoints by honouring and integrating the contributions of each person.
>
> Consensus is not an easy meeting form nor a panacea for all meeting ills, and its pathway is paved with many misconceptions which can result in unsatisfying experiences.
>
> It is a decision-making technique in which all members of a group actively participate in reaching unity. It is based on the belief that each person

Dining hall in a Bruderhof community

The Bruderhof community is the closet thing I've come to kibbutz when it comes to everyday living, working, eating, family life and economics. We visited two Bruderhof villages in upstate New York in 2004, already some years after we had left Kibbutz Gezer. Eating in their dining hall was like a flashback to old times on the kibbutz: nearly four hundred people all eating together, the sounds, the sights, the feelings. It was highly nostalgic!

Conversations were mostly around the tables with your immediate neighbours, but they had one interesting addition. A cordless microphone was handed around to anyone who had something to say. As soon as they began talking through the loudspeakers, everyone would fall silent and listen. Anything could be said, it was kept short each time, but it was something that the person speaking felt to be on their mind. Quite unexpectedly, I was handed the mike and asked to say something, and managed to stammer out a few unprepared words of thanks for being there. It all sounds a bit intrusive, the way I have described it, but the feeling at the time was very comfortable. It kept a sense of togetherness, and the periods in between announcements were used to comment what was said, and other general talk.

The feeling of togetherness was very strong.

> holds some part of the truth of the group; no one person holds it all. There is no voting. It eliminates majorities and minorities and avoids the potential polarization of yes and no factions created by voting. It seeks to reach the highest and best solution, rather than compromising to a middle ground or settling for the lowest common denominator. The process itself can build trust and create a spirit of community within the group.

During those first formative months and years, a community would be well advised to look consciously at the process of decision-making. If training is needed, then is the time to establish a course for the community, and most importantly, for new people joining. It would be sensible to build in plenty of opportunity for informal togetherness. Meals are an obvious choice, but just as good are shared break times, or shared work. The point is to get people to talk together. Encourage people who want to bring up issues to talk it through with as many others as they can before the issue comes up for formal discussion or decision-making.

Consensus decision-making and conflict resolution will be covered more in Chapter 8. If you establish good practice early on, and avoid conflicts, you can skip those bits.

3. Not Counting the Hours

Members of new projects, especially when they live and work together, tend to work long hours, fuelled by initial enthusiasm. Any project which engages people, which lives off their enthusiasm and motivation, will get them to put in that extra bit. Intentional communities are built on ideals of creating some kind of social improvement. Equally, there are many groups and organizations which also aspire to improvements of one kind or another, and most of these will be full of people putting in long hours of hard work, happily and enthusiastically. At least at the beginning. When we come to focus on eco-communities, it's clear that because we live in such close proximity, it gets even easier to put in extra time. Being even more specific, and focussing on communities that pool their income, this trend is accentuated even more. In our early years on kibbutz, no record was kept of hours worked, and barely of days worked even. Here in Camphill we still have no time-keeping of a formal kind. When you operate at this level of trust, the extent of self-exploitation (which we call overtime anywhere else), is extremely high. Given that you have financial security unrelated to the job in hand, it's amazing how much can be done by purely voluntary work. The satisfaction of getting things done is often enough of a reward in itself.

When we look a little closer at this, it's actually nothing unusual or remarkable. Many people like to potter about the house, fixing or improving things. Others like gardening, for flowers or for vegetables. The amount of voluntary work, often done for charitable institutions, is staggeringly high, an enormous statistic that is usually overlooked by official statistics.

A few months ago I talked to a number of ex-Camphill co-workers here in Norway, to find out what aspects of Camphill they had taken with them into their subsequent careers. What had they gained from living in community? This was not a random sample; they had all gone on to become Steiner schoolteachers. The object was to see what, if anything, had been learnt from community living and taken into the schools afterwards.

Working together for the common good

Working hard together can really forge bonds between people. I have often found that when working together with someone on a common task, that we get to know each other in a completely different way than if we just talk. Doing both can be really good. Once I was pruning vines with a group of people, day after day, in a large vineyard, and once I got the hang of the pruning, there was not much else to do than work along the rows, the actual pruning went pretty much automatically. I was getting to know another of the pruners, he seemed really interesting, and we decided to trade life stories. The next day he had all day to tell me his biography, the following day I would have the whole day to tell him mine. It was an amazing exercise in learning about someone else. The work got done, and we both felt that we had gained a new friend.

Within community, working together, seeing tasks get done, seeing things being built and progress being made, really creates that feeling of togetherness which is one of the building blocks of the collective. Confronting problems, overcoming obstacles, working out solutions can all be so much easier when there are people doing this together. What one person finds hard, another might find easier, when one person feels that he or she cannot get any further, another will find an opening and forge ahead. Afterwards, when the job is over, and you can see the results of your efforts, the feelings of comradeship really flourish.

Volunteer building trip to Russia. Courtesy of Rolf Jacobsen, Norway (see also p. 131)

Several of them commented on the experience of not working to a time limit, of putting in all their energy because the work itself was worthwhile. In the Camphill villages the boundary between what is work and what is private is not so clear; in short, one works pretty hard. It was generally agreed between those I talked to that this was a good training for life as a teacher.

One teacher commented: 'I learnt to work, and to take responsibility!' Another pointed out: 'The workload in the village is very big. Camphill helps to develop willpower, you are busy the whole time, every evening. You are engaged in everything.'

A teacher who had grown up in Camphill made the following statement: 'What I got from the life in the village was the feeling that the fellowship where I worked was US, that we were not just employees, but part of a living organism. This makes it easier for me to help carry the load, even though I don't get paid overtime for extra hours put in. It is very important that we don't measure the economic side of everything that we do.'

After spending a year as a volunteer at a Camphill village in another country, a friend of mine went on to study agriculture, eventually teaching and becoming the head of the first agricultural college in Norway to take organic agriculture seriously. She commented that even though this was a college, not a village community, and that everyone was an employee, there was much more of a feeling of community when compared to other colleges. Everyone did what needed doing, and didn't count the hours, and that this came naturally to her, reinforced by her Camphill experience.

Margarete van den Brink studied organizational structures in Holland and comments in her book *Transforming People and Organizations:*

> Such dedication can also be found in other ideal based organizations. A woman who exchanged her job in a commercial organization for that of campaign-leader with Greenpeace, says: 'Here you do not think about noting down quarters of an hour. You work far too long hours anyway. You work from your own motivation. I do not talk about my work with Greenpeace, but about my life with Greenpeace.

This is what people will remember from the early days, from the campaign trail to set up a new venture. This is one of the sources of myths and legends specific to each community. All those who were involved in the wheel rolling of Cartwheel in that long hot summer of 1980 will remember the evenings spent talking, the hours on the road, and the meetings with others on the way. Even though the project itself subsequently collapsed, the wheel rolling itself became part of the personal biography of many individuals.

Kibbutz Gezer was never a religious community, though there were always religious people living there. The culture is decidedly Jewish, with celebrations marking the Jewish calendar, and with Hebrew the main language. At about the same time as the Christian Easter, in the spring, Passover is one of the great Jewish festivals, the high point consisting of a celebration dinner, the Seder, during which the Israelites' exodus from Egypt is recounted. This lasts several hours, with readings, discussions, presentations and drinking of at least four glasses of wine. We happened to be visiting Gezer just at this time, in its early days, in the spring of its third year. We were gathered together in the dining hall, perhaps a hundred people all together, getting into our first glass of wine, when one of the members in charge of fields came in to announce that there was rain being forecast and that there were thousands of bales of hay scattered about the fields, having been cut, dried and baled that day: 'I don't want to interrupt the Seder, but I would invite anyone who wants to help to come back to the dining room afterwards in their work clothes, and we'll get what we can under cover before the rain comes.'

The Seder continued, with most people drinking their four glasses of wine, and several of us being much more generous. After the dishes were cleared, and the place tidied up, people began drifting back in their work clothes, more and more of them. It seemed as if everyone came back. So many that a kitchen crew was assembled to make hot coffee and soup, and the electricians zipped off to their workshop to rig up arc lights around the barns. The rest of us divided into work teams, each one with a tractor and a wagon, and off we drove into the fields to load up bales.

Each time we returned with a loaded trailer, we were greeted by steaming cups of soup and coffee, and lit up barns to stack into. Enthusiasm, fun, and dedication pulsed through the crowds. When the dawn came, most of the bales were in, and as we stumbled home, exhausted, we met the morning milkers on their way to the dairy.

This story from Gezer will be recognized and shared by those who were there. It's a bond between us, a shared history, a consciousness of the group. As such, much higher value than the hay bales. It's difficult to estimate the value of these community biographies. My intuition tells me that the strength of a community is only partly based on material and measurable things like financial success or building mass. More and more, as I live in community and observe community, I see that there is something beyond, behind or above the material which gives the community its soul, its essential life force, its individuality. Part of this individuality is the shared and remembered history. It may often feel like endless hours of overtime, but it contains moments that will live on in our consciousness.

Another big advantage of working hard regardless of time is that things *really* get done. It's amazing how much can be carried out, and what improvements can be made, in a short time. This gives momentum and stimulates motivation, more ingredients for fun community living.

It seems that every silver lining has a rain cloud attached, and there are also risks attached to working long hours. Resentment can build up between those who can carry a big workload, and those that do less. When nothing is measured, and when people do things out of their own enthusiasm, some people end up doing more than others, or so it seems to them. Egoism is always there ready to poison us when selflessness slips from our consciousness. How come that other person ...? And so on. There is no recipe against this other than a deliberate cultivation of selflessness on the part of each individual. This is one of the core characteristics of communal living, one which attracts some people, and really repels others. In the very messy and painful conflicts which have been known to develop, we might later experience individuals billing the group for overtime done during this initial boost of enthusiasm. Of course we hope it doesn't get to that, but the resentment factor is a built-in trap we would do well to be aware of.

Bringing in the bales

Occasional events stand out in the early years of every community. In the case of Kibbutz Gezer, for me the bringing in of the bales on the night of the Passover meal was one such event. The experience of virtually the whole community rising to the occasion, working together, disregarding free time, and focussing on a common task for the common good, has since become one of my community stories.

This picture was taken many years later, nearly twenty years later. These are the children of the members who were asked to come out after the Passover meal. Same fields, same type of bales. If you ask people today, there are other versions of the story; Manipulation on the part of canny farm managers, or panic on the part of inept managers. I prefer the 'great stories of the past' version. These stories play an important part in building up the community's identity, they become part of the community biography, and live on.

Another risk area might be called 'owning the project by our enthusiasm.' An activity becomes totally identified with one person, and others don't see their own creative role so clearly. This was what happened in the anarchist commune in East Anglia mentioned in Chapter 1. This is a feature very specific to the early stages of a community. Later, it becomes perfectly acceptable and is called specialization.

For many people, the satisfaction that they get out of life is related to how much they put in. This is a human characteristic which is well worth spending some time thinking about, both individually and as communities. Within communal living, and especially as we see it in the explosion of ecovillage activity over the last decade, it has had tremendous impact. Working towards higher ideals, together with others, not only creates real change in the world, but can also create real change within the individual.

4. Not Enough Money

If you want to make a lot of money for your investment, communities are not really a good bet. With the primary aim of changing society, or at least creating the good life, increasing the value of investment made is far down on the list of priorities, if indeed it figures at all. There's an old joke that's been circulating in the Middle East for years:

Question: 'How can you make a small fortune in Israel?'
Answer: 'Come with a large fortune!'

That might be equally true of intentional community. Most of the activists who get into this are not famous for their fortunes, either large or small. In fact, the personal wealth of a single person can often lead to very difficult social situations. A collective debt to an outsider is easier to handle than the same debt to an insider. When one person buys land and buildings, it is often difficult to give this into a shared situation, and there will nearly always be a tendency for that person to dominate the group, using their wealth to bend the group one way or another. This builds up resentment in others, and is a sure way to spark off conflict.

However, the lack of ready capital may not be such a big handicap. In Permaculture planning we often talk about finding the solution within the problem. If a community needs capital, instead of becoming paralysed by its poverty, it can turn this into an active search, a common task, and spend some of its energy looking for support. Financial capital is only one of several ingredients for the project to succeed. Creative brainstorming sessions may give a much wider perspective.

Take some time to concentrate upon what resources you have in the community. Write down a list of everything that comes to mind. You'll be amazed at how wide these resources can be. When the energy level gets back up, make another list of which resources you need to make the community a reality. A really interesting task following this is to match up these two lists, to see how far you have come already. A third list can be made of which people, organizations and official bodies can be possible support for the project. Matching the first two lists with this third one will give you an action plan, to which you can add who should approach which support groups.

It may well be that this puts things in perspective. The lack of capital can be seen in its wider context, and concrete tasks can be assigned to finding the right resources. It may also become clear that financial help can be sought from a variety of sources. A request from each source should not be too big, and once one or two prestigious sponsors have been located, it often becomes easier to raise support from further sources. For example, a modest housing loan from a recognized bank may itself include free consultation services for further funding assistance. If you can build a feeling of partnership into this relationship, you will gain valuable support for a long time to come. Support organizations live off giving support, and gain prestige from being involved in successful ventures. Make it one of your aims to become a flagship project for the organizations supporting you. If you get media coverage, mention them as partners, and send them relevant press clippings.

Let's go back to Cartwheel and learn more from their experience. Even though nothing finally came of it, the action of rolling an enormous six-foot cartwheel by hand around England brought the group into contact with thousands of people. The idea of a community, aiming to be self-reliant and ecological, fell on fertile ground, especially in the economically depressed areas of northern England. With a history of unemployment and exploitation, local councils in the north were genuinely interested in Cartwheel's ideas. In the industrial areas of Lancashire, several discussions were held with local councils who wanted to know more. Had Cartwheel been able to go back to talking to them after the wheel rolling, there might well have emerged some concrete projects. Sheffield City Council also became very interested, and there was a meeting with a group of councillors and city planners, who offered a site, and asked for an outline plan as the next step.

The possibilities were there, the offers were made. The fact that Cartwheel was not up to carrying through negotiations with the Lancashire councils, or to present Sheffield with an outline draft plan, was Cartwheel's problem, its own inability to function.

Ecological banking

Ethical and ecological banking emerged seriously over twenty years ago, and went through a number of stages. I remember the Ecological Building Society in England being set up to help small alternative projects, by giving savers somewhere to put their money where it would be used for environmentally friendly building projects.

Combining the building society idea with the banking idea created the possibility of positive investment, the lender choosing either an area or even a specific project that they wanted to support.

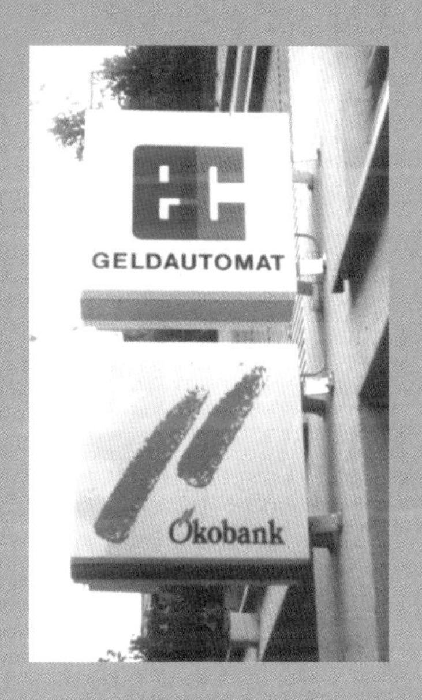

I took this picture of an ecological bank sign in Freiburg, in Germany, in the mid-1990s. Inside the bank I found notice boards with information on many ecological projects. There were advertisements for ecological products, explanations for different projects, communities trying to attract new members, advertisements for alternative schools and kindergartens. This was a real meeting place for alternative minded people. One of the fundamental Permaculture principles is multi-functionalism, and here the bank had become something much more than just a place you can leave your money safely.

Since then the idea of ethical banking has been taken on board by many larger banks, which offer their customers some kind of assurance that their savings won't be used for investments in companies that do really bad things: armaments, human rights violations or heavy duty pollution.

En eco-community will eventually turn over a good deal of money when it gets well established. To do this through a proactive ethical bank will mean an increased turnover for that bank, and will give an extra boost to the ethical banking sector. We would do well to put our money where our mouth is.

In Norway, the Kilden ecovillage group had the opportunity to compare projects in very different locations. Kilden was set up in the 1990s with the idea of creating an ecovillage near Oslo. A number of sites were considered, and by the time I got involved with them in 2001, there were three sites being explored. One was being offered to them by the local municipality of Hurdal, an hour north of Oslo. The second was in the commuter suburb of Nesodden to the south of Oslo, while the third was based on an existing farm to the east of Oslo.

The initiative from Hurdal was from an area which was finding it hard to attract people, and where there was a lack of development. Some of the more creative local politicians and administrators saw in Kilden the possibility of attracting a group of highly motivated younger people who would set in motion a project which would attract both people and capital to the area. Today Hurdal is a thriving eco-community, with a number of families on site, having built their own first cluster of houses. More families are joining every year. The project at Nesodden was squashed by the local authority, citing planning regulations, but later it transpired that it was mostly due to complicated local political in-fighting. Interestingly this project had the strongest economic base, and was being developed by a group of experienced architects and planners, while the Hurdal group at the beginning had very little apart from youthful enthusiasm. The third project, the farm to the east of Oslo, just never materialized.

There seems to be a pattern emerging here, of the periphery versus the centre. In and around large cities there is pressure on land, it is expensive and there are many vested interests attached to it. In more outlying and depressed areas the pressures are different, they are tied up with the issue of attracting people, enthusiasm and money. When a group presents itself with an idea of establishing something economically viable, which will attract more people, this becomes attractive to such outlying host authorities. Just as in retailing, location is the key to success.

A similar pattern can be seen in rural Spain. Some years ago a friend of mine was looking into the possibility of setting up an ecological village of artists in the rural area inland from Barcelona. When he came back from an initial exploratory visit

Ecovillage planning group

It seems to take a long time, and considerable effort, to set up a new community. When I moved to Norway and got involved in the Kilden group, they had already been meeting and talking for several years. These pictures come from the time when the ecovillage at Hurdal was still at the idea stage. The first is from a meeting of the umbrella planning group at one member's house in 2002. The second picture is from a much larger meeting at a potential site which later turned out to be non-viable.

Only in 2006, nearly a decade after the Kilden group was created, has Hurdal ecovillage really become established, with building plans approved, a viable financial structure and a serious, committed, growing membership.

he told me that he had been offered whole deserted villages by the authorities, who were even prepared to help him seek funding from the European Union.

When we come to the issue of planning and building regulations, I have found a similar pattern of great differences between where places are located. When I lived on Kibbutz Gezer, I was involved in the building group, and planning permission was needed for every little thing we did, even the building of a veranda or back porch. During that time I visited The Farm in Tennessee, and was very impressed by their experiments in strawbale building, and other ecological techniques. I asked how they got around the building regulations. Albert Bates, the director of the Ecovillage Training Centre there answered that there were none. They were too far out in the country, and you could build anything you liked. Total freedom!

This polarity between the centre and the periphery makes it harder for communities to establish themselves in or around large cities, where land prices are inflated by demand. The only exceptions seem to be sink areas within cities. Slums, depressed areas and former industrial sites may present themselves as socially or financially peripheral, even though they might be right in the geographical centre. There are plenty of examples of thriving urban ecovillage type projects, many of them with very little financial clout, at least at the beginning.

Setting up systems to borrow, manage and store your money is quite crucial. The systems that you install at the beginning will often determine how things turn out in the future. Creating relationships with banks and loan institutions can make or break your financial viability. Here in Norway the Camphill movement has been lucky enough to establish a close working relationship with Cultura, an ethical bank inspired by anthroposophy, Rudolf Steiner's framework of thinking. It's a small bank, and we know many of the people working there personally. When you call them up, and they recognize your voice on the phone, this already makes life easier. Even more important, we speak the same ideological language, we share aspirations, and we trust the bank. Working together with them, we have been able to set up a pension fund for the Camphill movement in this country, some of which is now available to us as a source of loans.

Hurdal ecovillage on the map

My involvement with Kilden (The Source) Ecovillage Trust in Norway dates from just after a sub-group within the trust had come up with the idea of starting a community at Hurdal. There were other projects also being followed up, and at times the Hurdal sub-group seemed highly speculative. Sometimes it seemed there was little progress, all kinds of bureaucracy got in the way; there were financial problems, and the group waxed and waned. Kilden was during this period pursuing two other projects, both of which turned out not to be viable, and eventually the Hurdal group got themselves established, set up their own legal entity based on a co-operative model, and broke away from Kilden.

Since then, I have kept in touch with Hurdal, visiting once or twice a year, and arranging for Permaculture meetings to be held there a couple of times. It had the feel of a typical pioneering community, just hanging on by the skin of its teeth.

At their Open Day in the summer of 2006 (see below), I noticed a qualitative change in the atmosphere of the community. Their building plans had been approved, an economic structure was pretty much in place, there were several new members, and more on a waiting list. The place was dominated by families with young children. For the first time I felt it was an established community.

The transition from pioneer to established, from youth to maturity, had been accomplished.

The ecovillage at Hurdal managed to set up a good working relationship with the same bank, and also with the 'House Bank,' an organization providing mortgages for house buyers. This gave them working partners within the financial world, and the ecovillage was successful in presenting themselves as a serious concern, and were in turn taken seriously.

If we really want to turn this problem (not enough money) into a solution, we find that by being forced to work together with outside financial institutions, we can spread the idea of small-scale ecological future communities into areas where we might otherwise never get to. This in itself is valuable; it's a part of the basic message that ecovillages are about. At the same time, it provides a stable foundation for future growth. Later, when we come to look at how communities mature and grow older, we'll find that if good foundations are laid at the beginning, this ageing process will be much smoother, and that the ecovillage will have better prospects for a long and healthy life.

5. Open the Door or Close the Gate

The tightly knit group that works hard together to establish a new community can often be difficult for new members to penetrate. But unless the aim is to remain a small intimate group there will be a need to grow and take in new members.

This issue of membership requires delicate balance. It is essential to have a core group that sets ideals and strategies, otherwise there will be a lot of floundering while attempting to get established. But how can this core group open itself up to new members, essential for continued growth? The challenge is how to attract these new people and how to give them room to make themselves part of the group, while still retaining core values.

The answer seems to lie in a clear membership process. As long ago as 1982, Sarah Eno and Dave Treanor wrote the *Collective Housing Handbook,* one of the first real 'how-to' handbooks for communities. In the chapter devoted to membership issues they state clearly: '... you will need to check out a prospective member — perhaps with several visits and a trial or probationary period before they are finally accepted.'

They go on to describe the process they had established at Laurieston Hall in Scotland. After a few initial visits, the prospective member asks to stay for a trial period, usually between six weeks and three months. If successful, and there are no objections to them staying, they then attend a Joining Meeting, which was highly structured, ending in a decision being made by the community. Laurieston Hall has been a stable community, with members joining and leaving, for well over thirty years now. I visited the place twice, the first time in about 1980, when I attended a woodstove building course, and once again briefly a quarter of a century later. During my second visit I was impressed by how much the place had retained the communal flavour which I remembered. Even though there had been lots of changes, and many individuals had come and gone, the community had created an identity which I recognized. Of course there are many other factors which go to create that unique identity which a community can attain, but the process of membership remains one of the important ones.

Kibbutz Gezer had a much longer process through which Ruth and I made our way in the mid-1980s. First there was an initial visit to get to know people and make a clear application. This was followed by a six month trial period, culminating in the community voting us in as candidates for membership. Candidacy lasted a full year, which could be extended if one was away, or if one of the couple became pregnant. During this period we enjoyed full rights of membership, received the same pocket money and other paid expenses, attended community meetings and could sit on some committees, but we had no voting rights when it came to community decisions. The process was monitored by a 'Joining Committee' which gave us a contact person with whom we had regular meetings to check that everything was going well. When it came to the final vote, it was usually quite an event, there was high attendance at the general meeting, and it was tradition that afterwards there was a party at the new members' house. In our case our candidacy was prolonged because Ruth became pregnant, and the vote was taken on the night that our second child was born. Neither of us was on the kibbutz, but our welcome back with a new child was very warm.

Candidates for membership were not just individuals and young families, groups also came. They were organized through the kibbutz movement, some came from abroad, others were young Israelis. Their initial application was as a group, but each person had to go through the candidacy and membership process as an individual. Not everyone stayed. A group of Brazilians came just before we arrived; several of them became members, but within a decade there was not a single one of them left on Gezer.

Even this long process is by no means foolproof. There are so many other factors bearing in on membership issues. Negative aspects are sometimes overlooked in prospective members because other features are more attractive. The community really needs an electrician, and the aspiring member is a qualified electrician with lots of experience. So let's vote him in even if he is a jerk. A couple applying consist of a highly desirable and qualified farmer, while his wife is intolerable. The community desperately needs someone to run the farm. Do we shoot ourselves in the foot and reject them? The hopeful member is not very likable but the brother of a respected long term member. How can we vote no?

Kibbutz members celebrating a festival

When we became members of Kibbutz Gezer in 1986, the community was growing rapidly, just topping one hundred full members, not counting those in the process, volunteers and children. There were over two hundred people actually living there in total. I remember being told that it was one of the most rapidly growing kibbutz communities in the country. There were really no limits to growth, we were applying for loans to build a new dining hall, and were planning for a four to six hundred seating capacity. A new neighbourhood had just been built, another one was being planned, and additional income generating enterprises were being contemplated.

I look back to this time as the golden age of the kibbutz, there was a real feeling of vitality, a strong radical ideology, and life was generally fun. This picture was taken during that period, at the celebration of one of the Jewish festivals, events which were attended by most of the people, with the kids often playing a central role by presenting some song or play or other show. Celebrations helped to cement the community together, and often ended with big, long and loud rock parties, where the drink flowed freely.

People can also hide their true aspirations for a long time. On the kibbutz I thought our one to two year procedure would reveal most people in their true colours before they finally came up for the vote. Ruth and I eventually left the kibbutz during a period when the community was going through a strong decollectivization process. One member who had lived there for several years told me that the only reason he had joined was that he saw the community ripe for disintegration, and that he figured that when the collective finally split up, he would come out of it privately owning a valuable real estate property. I was devastated. He had seemed such a nice guy when he joined.

In Cartwheel the question of membership was discussed at great length, but no formal structure was ever set up. This had the effect of giving newcomers as much voice as everyone else immediately, even though they were not familiar with previous discussions and decisions. In practice this had the effect of bringing the group back to the beginning very often. Much of the work of the general meetings during the first year was concentrated on hammering out a vision of principles, but this took much longer than it should have done because new members continually arrived, and needed to go back over issues that had already been discussed.

Some communities may not want new members at all. It may be that they want to shut themselves off from the outside world, an attitude that I find hard to reconcile with the broader span of ecovillage ideals, one of which is to effect social change. I can't see how a community that doesn't want to interact with the rest of the world can influence it in any way. But more members need more space, and if the size of the house or the plot that a community settles itself in is small, there's not much that can be done.

Parsonage Farm is a small community, set up in the early 1970s in East Anglia in England. A close friend of mine moved there a few years after it was founded, and I visited several times, even staying there for a month in the early 1990s. For a couple of decades it remained a very stable community, the core group of four or five people held it together, and it felt very much like an extended family. The house itself could not contain more than about ten to fifteen people, and it remains more or less the same size to this day, even though there have been changes in membership. In these circumstances, with a

Parsonage Farm Community

Not all communities want to grow. There may be many reasons for limiting growth, among them the available space and financial resources.

Parsonage Farm was set up in the early 1970s in East Anglia in England in a small rural village. I had a close friend who lived there, and kept in contact with the community, visiting and even staying there for a while. It was pretty stable, with a core group who lived there for a couple of decades, but it always remained small, ten to fifteen people at the most. In many ways it resembled a large family rather than an organized community. With such a small membership, there was less need of formal structures. Many decisions could be taken round the dinner table.

The picture below was taken in 1991 when the community was already twenty years old. It was a small group, who ate around the same table every evening. There was a pretty relaxed atmosphere, but of course with all the inner tensions that one might expect from a group of people living closely together.

small family sized community, hard rules and procedures are not always so appropriate.

Coming back to the subject of village sized communities, which really need to be more than a dozen people if they want to qualify for such a title, it is appropriate to have some rules or guidelines. Diana Leafe Christian in *Creating a Life Together* makes a very similar case to that made by Sarah and Dave of Laurieston twenty years earlier: 'NOT having a membership selection process can be a heartbreaking source of structural conflict later on.'

From her extensive experience of watching community, as the editor of *Communities Magazine,* and one of the members of Earthaven ecovillage, she goes on to elaborate:

> Many people don't think it's 'community' unless the group is inclusive and open and anyone can join. Doesn't community mean offering a more accepting, inclusive culture than mainstream society? Most experienced communitarians would reply that not having criteria for new members — admission standards, if you will — is simply an invitation for emotionally dysfunctional people to arrive.

It can't be stated more clearly.

Given that we want to grow, having in place a clear vision statement and a clear candidacy process, what might be the optimum size for an eco-community? This question will come up in the early days of a group, and there is often a spectrum from those who want a small intimate group to those who have ambitions to form a large community.

The kibbutz movement has often been referred to as one of the largest and most highly profiled collective movements of the twentieth century. With well over one hundred thousand residents, and nearly a century of development, it certainly has a wealth of experience to give food for thought about how to organize an income-sharing, democratic community. The debate about size raged within the movement right from the start, and took several years to settle down. The very first kibbutz communities we can identify, with Degania founded in 1911, called themselves 'intimate groups' and wanted to limit themselves to a dozen or two

Large group dynamics

The dynamics within a large community are much more complex than those of a smaller one. First of all there are many overlapping circles of contact. The people who live immediately around you are visible all the time, and give rise to many short chance meetings during the day. The group that you work with provides daily social contact, and a need to organize the workplace with all that this implies in terms of discussion and decision. Your close friends are not usually your immediate neighbours or your colleagues at work, and provide yet another social context. Then there are other circles, depending upon how the community is organized. If you have communal child care and if the community is large enough to have its own school, there will be parents groups, for example.

Putting together all these circles of interaction will create a complex social ecology within the community, opportunities to interact with many different members in different ways. This helps to create the social glue which holds the community together from day to day, week to week. If the community is large enough, it will satisfy most of your needs for social interaction.

members. Growth was to be achieved by the setting up of new communities. After the First World War, the arrival of highly politically motivated Jews from Russia gave a whole new twist to the kibbutz movement. These were people who had visions of large communities, a hundred, a thousand or even more. At this point, Palestine was administered by a British Mandate, and most of the Jews living there had ambitions to found an independent state, the state of Israel. Some of the political kibbutz members talked about a super kibbutz, the whole state as one enormous collective, everyone income-sharing, no private property, and collective ownership of the means of production and the agricultural land. These must have been heady days indeed, and I can only imagine how heated must have been the discussions between them, and the original kibbutz founders and their ideals of the 'intimate group.'

Once the extremes have been mapped out, nature often has a way of achieving a balance between the two, and coming to a sensible compromise. During my period of involvement in the kibbutz, it seemed to me that a membership of a few hundred was the norm. Gezer, where we lived, hovered around the one hundred mark for a long time, and was considered small. A neighbouring kibbutz, where we had a lot of friends and visited frequently, had around four hundred members, and was considered to be average size. It has to be borne in mind that here we talk about members who had been through the candidacy process. In addition there were children, volunteers, candidates and other visiting groups and individuals. As a rule of thumb I used to reckon that the number of actual people on a kibbutz was about twice that of the formal members.

Interestingly enough, this seems to correspond with other people who have given thought to what might be an optimum size for a village. Charles Handy, writing about management in the late 1970s, estimated that the 'organizational village' (his term for a corporate company!) should be a maximum of five hundred people, above that number, anonymity sets in and not everyone can know each other. Handy wrote his book a long time before ecovillages were thought of as such, and as a management consultant and guru, there is little evidence he had much contact with alternative communities.

In the early 1990s, when Ross and Hildur Jackson asked Diane and Robert Gilman to write a report on ecovillages, the size of such a 'human scale settlement' was suggested to be between fifty and five hundred members. Whether they had read Charles Handy or not I have no idea, but it seems to me, both from reading, and from personal experience, that their estimate is pretty correct. A community that reaches this size will long before have broken up into subgroups, circles of friends, work teams, or neighbourly groups, which seem to be most comfortable at around a dozen individuals.

It's well known that most armies use the platoon of between ten and twenty soldiers as their basic operating unit in the field. This seems to be a number which works well together under stressful conditions. Twelve is a magic number. Moses headed out into the desert leading the 'Children of Israel' divided up into twelve tribes, and Jesus is well known for having twelve disciples. He really changed the world with quite a modest group! For those who are fascinated by these patterns, there are twelve months of the year, twelve signs of the zodiac, and when I grew up in London, there were twelve pence in the shilling. I rest my case: the intimate group seems to be in this size group. I would call this a family size. Above that, a community can grow to several hundred with a socially comfortable number of people knowing each other.

What happens if an eco-community continues to grow? In the current network of GEN ecovillages, Auroville in southern India is well over the thousand mark, and Damanhur in northern Italy is close to a thousand already, and there are plenty of other ambitious communities out there. My personal experience is that anonymity becomes more pronounced as the community gets larger. This is not unwelcome to many people, who are unhappy with the lack of privacy which can result from living too close to a small group of people. In the end it's a matter of taste: some folks prefer the intimate group, others like the wider spectrum of social interaction.

Just to illustrate where this can get us I will end this chapter with another anecdote from my kibbutz days. Many years ago we lived for a year on a kibbutz which then had about five hundred members. One of our friends was invited to a neighbouring kibbutz for a wedding. This was kibbutz Givat Brenner, at that time, and I believe still, the largest kibbutz in Israel, over a thousand members, and two thousand people living there all told. They drove over, came through the gate, and stopped a passing kibbutznik to ask where the wedding was being celebrated. 'Oh, there's a wedding?' came the answer.

Small, intimate group

A small, intimate group will rarely satisfy all your needs for social interaction. Many of the communes I knew in the 1970s in Britain were small, often extended family-sized groups of between five and fifteen people. If successful, in the sense of being able to carry on over time without too many debilitating conflicts, they would provide a base and a safe haven, giving the members a secure home when they ventured out to do other things. Generally it's impossible for such a small group to provide an income for its members, schooling for the children, or other social services, so these have to be found elsewhere. This means a great deal of interaction with the outside world, which in turn will bring many social circles into the orbit of the small group.

There is a great deal of discussion within community over the issue of size. The Kibbutz movement began with the 'intimate group' and during the first years of the movement, just before and after the First World War, the ideal size for a kibbutz community was fifteen to twenty members. It was only when refugees began arriving from Eastern Europe, enthused with Marxism and Socialism, that the concept of the larger kibbutz community gained acceptance.

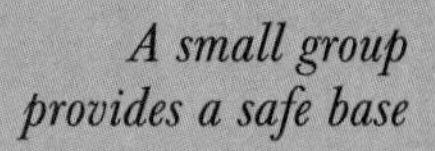

A small group provides a safe base

A larger community can organize effective work groups during the harvest

There is a lot of wisdom in the establishment of the larger community, with a variety of living options. A place where it is quite acceptable to live on one's own, to create a family-sized living unit, or a special interest group that lives together, or larger co-housing groups, all within the umbrella of a single alternative community.

We human beings are social creatures who find their base within the family-sized group. It is one of the challenges of a larger community to create the space for this size of group to establish itself, and to give the individual a springboard to interact with larger groups.

Our friends were shocked. In our, much smaller kibbutz, a wedding was something that engaged the whole community for weeks beforehand, food was prepared, entertainment laid on, speeches and sketches prepared, time was taken off from work, and everyone was somehow involved. It's a trade-off; you can't easily have a quiet wedding in a small community.

6. Flexibility and Multi-Use of Buildings

The first time I was invited to dinner at Ole Bull House in Vidaråsen Camphill village in Norway, I wondered why the dining part of their large living room was higher than the rest of the room. The food was brought in, we sat and ate, about ten or more people round the table, and afterwards, when everything had been cleared away, we settled in to armchairs two steps down in the split level living room for coffee and a chance to get to know each other better.

Vidaråsen was founded in 1966, in old run-down farm buildings, and year by year new houses were built and the community gradually grew. By the time Ole Bull House was being planned, there was a need for a space where the whole village could gather, and put on a show, hold a small concert, and celebrate things together. It was decided to give the house a slightly larger living room than usual, and raise one section of it to act as a kind of stage.

For several years, the Ole Bull House living room was the village hall. But the story doesn't end there. Ole Bull could not cope with the growing community, and eventually became too small. At that point, funding for a large, purpose built hall was not available. So it was decided to build a small hall, of a very simple design, which could later be turned into a workshop.

When we arrived at Vidaråsen in 1991, the main hall, the Kristoffer Hall, was only a couple of years old. The hall was large, impressive, with a real stage, large lobbies and coat-hanging areas, a good-sized kitchen, and even a small apartment for a caretaker or visitor. It could seat several hundred people very comfortably. Ruth worked for a while in the weavery, and every time I visited her I wondered why it had been built so large and square, and again with a raised section at one end. Then it was explained to me. This building had also done time as the village hall, and was turned into a workshop when the funding was available to build the Kristoffer Hall.

Joan Allen, in *Living Buildings,* cites many examples of re-use and conversion of buildings. About Vidaråsen she writes that it was she who suggested:

The village hall in your own living room

When Ole Bull house was built in the Camphill Village at Vidaråsen in Norway, there was no village hall, no gathering place for the whole community. So it was decided that the living room of the house would be larger than usual, and that a couple of steps up at one end would create a split level floor plan, which would do as a stage when so required. The house was built with funding raised by students selling candles on the streets of Oslo in the weeks before Christmas. For many years it served as a gathering place for the community, and many plays were performed there, as well as meetings and celebrations.

Today the upper level, which was the stage when it was being used as the village hall, is the dining room with easy access to the kitchen for food and crockery. The lower part is the living room area, and one corner has been enclosed as a small room, giving the space a more interesting, L-shaped, layout. These are simple, cheap, and easy to organize solutions. They imply wisdom and an understanding of how community develops over time, and would be an excellent example of Permaculture type flexible planning.

Ole Bull Hus Vidaråsen

The growing complexity of a mature community is often reflected in a large variety of activities. This signpost is from The Farm in Tennessee.

a simple, spacious workshop building which could be used as a temporary meeting place until such time as their final hall would eventually be built, after which the former could easily be converted into a full-time workshop. Vidaråsen adopted this suggestion, and the following year built Hans Nielsen Hauge Salen, a delightful structure which served for a time as an Assembly Hall and today is the weavery and doll workshop of the village.

The Camphill villages in Norway also had other ways of re-using buildings. Solborg Camphill village, where I now live, was originally founded in 1977, and within a few years this community also found that they needed a village hall. In the mid-1980s they discovered that a local chapel, built in 1888 and which had been out of use for many years, was about to be demolished to make way for road building. The Department of Roadworks allowed Solborg to take it down. It was documented carefully and stored. In 1987 foundations were built, a cellar was constructed, and the structure was rebuilt, with two small extensions being added. The whole process is described fully in Joan Allen's book.

The idea of re-using buildings, turning them from one kind of function to another, has become widespread throughout Europe over the last few decades. Often this has hit the media, as groups of radicals, anarchists, bums, drug addicts, layabouts (the name given largely depending upon which newspaper you read) took over old warehouses, disused office space or boarded up public buildings. Some of these groups were extremely short-lived. Sometimes they disintegrated because of internal dysfunction, more often because either the authorities, or gangs of hired thugs (sometimes hired by the authorities), threw out the squatters. This created good footage for the gutter press, another factor giving communes a bad name. More often, and usually without a mention in the same gutter press, these communities flourished, providing much needed space for the homeless, for battered wives and others in need. Very seldom was the environmental or eco-community factor taken into account when these activities were covered by the media, or debated by local authorities. This is such an excellent way of recycling buildings!

When we understand what an impact the embodied energy is of a building project, especially a large complex in an inner city situation, we begin to appreciate how important it is not to regard these structures as disposable, throwaway objects. Just demolishing a building creates tons of unusable garbage, another heavy load on the garbage heaps. Just the pollution from the machinery used to destroy and transport the stuff is immense, not to mention the dust created by tearing it down. If we can carefully re-tailor the building to a new need, we save enormous amounts of energy and avoid lots of pollution.

There is also a social factor. When a group comes together with such a focus, it can give that sense of togetherness and belonging that is often missing from the urban city centre. If they are allowed to work out their own plan, and given the right help to implement it, this can give otherwise hopeless people just the motivation needed to become useful members of their own small community.

Not just buildings can be re-used. Whole villages can be recycled. I have already mentioned my friend who was offered empty villages in Spain. The ecovillage at Torri Superiore in northern Italy, is a classic example, and was for a time the co-ordinating centre for the Global Ecovillage Network (GEN) in Europe. Originally built in the fourteenth century, the whole village consists of one building, with well over a hundred rooms. The Associazione Culturale Torri Superiore began restoration in 1989, moving into one room at a time, and rehabilitating the terraces surrounding the village on the steep slopes. This is a good example of recycling old buildings, this time with a vintage that is quite impressive. I just hope that some of the communities we are building today will have relevance five hundred years from now.

There are two more examples, both from Europe.

Lebensgarten ecovillage, one of the founders of GEN, and the European centre before Torri Superiore took over, was originally built by the Nazis in the 1930s as a factory and residential complex. Using slave labour, supplies of ammunition were produced for the war machine, until it was taken over by the British Army when the Second World War ended. When the Lebensgarten group bought the site in the early 1980s it had been abandoned for some time, and they themselves describe the site as being 'dilapidated.' Rehabilitation began in 1985 and today the place is a thriving, energetic ecovillage, one of our flagships.

ZEGG ecovillage, about sixty kilometres south of Berlin, and today the co-ordinating centre for GEN in Europe, has a similar history. Knowing it was built by the feared state police, *Stasi*, as a training camp, I can only shudder to think what kind of training went on there, when torture was one of the favourite methods for them to get information. When the Berlin Wall fell, and Germany was re-united, the time of the *Stasi* was over, and the place went up for sale. ZEGG, which stands for Zentrum fur Experimentelle Gesellschaftsgestaltung (Centre for Experimental Cultural Design) bought the site, and turned it into an ecovillage. Today, over a decade later, they are still rehabilitating old buildings there.

The village hall becomes a workplace

Anticipating the future use of buildings can be a tricky thing. Needs and situations change continuously in a community, people come and go, outside influences are often very difficult to predict, and new technology is often hard to anticipate.

This picture shows a good example of how future needs were correctly predicted, and how a modest budget was used wisely. At Vidaråsen Camphill village in Norway there was a real need for a larger meeting hall as the community grew and Ole Bull House became too small to accommodate everyone. Their architect at the time was Joan Allen, who worked for Camphill Architects in Scotland which served the architectural needs of Camphill worldwide. She suggested a building of modest dimensions, which would serve as a meeting hall until funds were available for a really large purpose-built theatre type building. The modest meeting hall could then be turned into a workshop.

With this in mind, Hans Nielsen Hauges Salen was built in 1972, and served as their meeting hall until the much larger Kristofferhallen was ready for use in 1986. When this building took over as the main meeting hall for the village, Hans Nielsen Hauges Salen was turned into a double workshop, the main section as a weavery, while the balcony was used as a doll workshop. The building continues to serve this function today, and is a large, spacious and comfortable place to work.

Weavery, Vidaråsen

Olavsal, Solborg

Mobile buildings

In Norway, the idea of re-using buildings is an old practice. Many of our traditional houses are built of logs laid horizontally, interlocking at the corners. This system readily lends itself to buildings being dismantled, moved to new sites and being reassembled.

Solborg Camphill Village was founded in 1977, and grew slowly around two focal points, an old farm that had been used as a school for deaf children, and a large boarding house which had been a school for socially handicapped youngsters. In 1985 a local chapel was about to be demolished to make space for a road widening scheme, and co-workers at Solborg were allowed to take the building down carefully, documenting the structure in detail, and remove it to Solborg. Foundations were prepared to receive the building, a cellar was dug, and two extensions were added, one to each end. The building was inaugurated in 1989, exactly 101 years after the structure was originally constructed as a chapel a few miles away.

It was renamed Olav Salen (the Olav Hall), in honour of the Viking king who had traditionally brought Christianity to Norway. It has become the main meeting hall, a concert venue for both musical and dramatic performances, and is used as a chapel for religious services on Sundays. Some years later, one side of the foundations was dug out and the cellar turned into a cafeteria with a kitchen area.

As I write this, discussions are taking place to design the area outside this cafeteria into an open-air gathering place and theatre, with an improved staircase, a water fountain incorporating a flowform, and a larger and better designed overhanging roof.

These two last examples, the Lebensgarten and ZEGG ecovillages, have an additional feature which I consider very important. Places which have such a heavy history can contain very negative connotations, negative energy, bad vibes. If left to themselves to crumble back into nature, they can become the foundation for the haunted house syndrome, a kind of spiritual Chernobyl. By re-inhabiting them, filling them with vibrant positive people who are socially engaged and creating a rich productive environment, those bad vibes are driven out. One of the principles of Permaculture planning is to concentrate on the worst affected area when you begin environmental rehabilitation. If you are open to the spiritual aspects, and want to improve the world we have inherited, what better place to begin than these?

Often things can be anticipated. In the ecovillage at Hurdal in Norway the initial cluster of small houses was built as quickly as possible, people were either stacked up like sardines in the only inhabitable building on site, or they were wintering in draughty, un-insulated mobile homes. Winters at Hurdal are not easy, lots of snow, and temperatures often dropping to –20^{0}C or below. Their longer term ambitions were clusters of larger family houses, further up in the forest. They knew it would take several years of planning work and negotiation with the authorities just to get permission for these, so they applied for a temporary building permit for this first cluster. Their hope is that as the larger clusters eventually get built, and the families can move out of the initial temporary housing that they can re-apply for permission to use this first cluster as workshops or summer accommodation for temporary volunteers.

I found another example of anticipating future needs on Kibbutz Gezer. The group that we joined originally moved on to the site in 1974, on to what had been a kibbutz, but had been abandoned. A dining hall was built, which served very well, but it didn't take long before it began to be clear that a larger building would be needed. When the laundry complex was built, next to the dining hall, it was built at the same level as the dining hall, and the planning group suggested that at some future date, when the dining hall would no longer be needed as such, the two buildings could easily be connected together, and the dining hall turned into a clothing store. This never happened, but it's a good example of creative, forward planning.

This concept of flexibility is important. In my experience of looking at communities as they grow and develop, I see that needs change, often fast and often drastically. It's important that building design does not lock the community into unchangeable uses. Counter to this runs the ambition to build things that are really custom built, reflecting the use of the building in its design. Architects inspired by anthroposophy often call themselves Goethean architects, and this idea that the building should reflect its use in its design is a central one. Throughout the history of architecture, ritual buildings, churches, temples and mosques, are great examples of this. Even within Goethean architecture, a cursory glance through Joan Allen's book will be enough to appreciate that the Camphill architects have evolved their own style of building sacred structures. The danger comes when this is transferred to secular buildings, especially working buildings. Workshops within Camphill can change a great deal. They are subject to change in leadership, in crew size, in community aspirations and the external demands of the marketplace if they are income-producing enterprises. I can cite numerous examples of the same space being used for weaving, carpentry, metalwork, office space and accommodation over a long period.

Here at Solborg we were planning a new workshop space a few years ago. Some of the co-workers suggested that we first decide which workshop should go in there, and then design the building to reflect this specific activity. Most of the planning group, including the architect, wanted to build an empty shell, which could then be fitted out for the workshop when finally the decision over which workshop it would be had been made. For totally unrelated reasons the planning process dragged on for several years, during which time several workshops came up for serious consideration, workshops as diverse as cheese-making, weaving, carpentry and even general storage.

The idea that you can tailor make a workshop is certainly attractive, but it should not lead to constriction of options at a later date. The initial planning of a *community* cannot anticipate

Lebensgarten

During the 1930s the Nazis built a munitions factory outside the village of Steyerberg in Northern Germany. This was a complete unit, not just the production buildings, but also housing for the slave workers and their Nazi masters. When the Second World War ended, the place was taken over by the British Army and served as a military base for many years. Finally the soldiers returned home, and the place gradually descended into dereliction until it was bought in the early 1980's by a group of idealistic people who wanted to turn it into a commune.

In Permaculture planning we often use a phrase: 'Turning a problem into a solution.' Clearly the Nazis were a problem, and even though it might be argued that the presence of the British Army helped to solve the Nazi problem, many of us feel that armies are generally a problem. A whole settlement in dereliction is certainly a problem. So here we have an ecovillage group taking on this multiple problem, and using the derelict buildings as a starting point, creating a solution.

The buildings were rehabilitated, as far as possible adding energy saving and energy catching features. Added insulation, solar panels, heat-catching greenhouses and the use of natural materials at all levels upgraded what the Nazis had originally built to become an efficient environmentally aware ecovillage.

On the spiritual plane, Lebensgarten countered the 'bad vibes' — left over from the past — by the presence of a group of Zen Buddhists who meet regularly, and by developing a conflict resolution business which has been used by alternative groups, industry and local government organizations.

On my visits to Lebensgarten, there has been a circle dance meeting every morning in the village square (see opposite), often preceded by a choice of half an hour of sacred singing or Zen meditation. My feeling is that the ghosts of the past have been driven away, replaced by a vibrant community, open to positive spiritual influences, and practising 'living lightly on the earth.'

the changing ideas, opportunities and conditions within the village as time goes on. Inevitably, things change, and the need for some types of buildings will be replaced by other needs. This has to be reflected in flexibility of housing design, and a willingness to change.

Our needs, both private and communal, are changing faster than ever. This requires a flexible approach to planning, and an openness to change.

Part 2

Maturity and Stability

Intentional community has been a feature of our western civilization for centuries, millennia even. We can go back to the beginnings of Christianity, even to the Essenes, and find people rejecting the mainstream society and setting up alternatives. Usually, at least up to our modern era, these communities were infused with a religious impulse, though often at variance with the established church.

Communities seem to have a life of their own, taking ideas and carrying them from generation to generation. Within our larger western society, I regard them as social organisms, which behave according to patterns that we can see and recognize. Rudolf Steiner saw something similar and writes in *Ideas for a New Europe:*

> ... whatever comes about in the course of human history is something living, and not a mechanism. Whatever is alive goes through growth and decay.

It is this pattern of growth and decay which I find fascinating, and I have been arranging my observations, comments and thoughts into three simple phases: youth maturity and old age. In the first part I looked at communities in their youthful phase, now I am going on to the second phase of maturity. This is the phase where communities settle down, gain a greater and hopefully stronger sense of identity. Friedrich Glasl, in *The Enterprise of the Future,* traces the development of businesses and calls his second phase 'the Differentiated Phase,' where formal structures and regulations become established. For him, this phase is dominated by system, order, logic, and control. It is much more businesslike and rational than the pioneering phase. Division of labour begins to be much more obvious.

Camphill co-workers Michael and Jane Luxford spent over two years travelling round the world of Camphill, and compiled the results of their conversations and observation in their book *A Sense for Community*. They hint at the idea of the Camphill movement having growth phases somewhat like a human being:

> On the Sunday between Ascension and Whitsun 2003 the Camphill movement celebrated its sixty-third birthday: 1 June 1940 – 1 June 2003. In his lectures on education, Rudolf Steiner spoke of the significance for human beings of the first three seven year periods in child development. This period covers the time between early childhood, childhood proper and adolescence up to early adulthood. He later went on to speak about the further stages of these seven-year phases, which continue throughout life, each having their own character and purpose.

In this second part, looking at communities in a more mature part of their lives, I don't want to lay down laws or draw up rules. I want to look for patterns and trends. I want to tell stories about communities, and retell observations and thoughts from others who have pondered how human groups develop. For those of you who live in community, or are planning an ecovillage, or are just working with a stable group of like-minded people, it should be interesting to observe some of the behaviour patterns of the group. Just as human beings change their behaviour as they grow, so also do human groups. Understanding those changes will help us to cope with the inevitable problems and conflicts which we encounter on the way. But this is not an exact science, neither will it foretell the future for your eco-community.

We can only really predict after the event, and that's good enough for me.

7. Procedures for New Members and the Rule of Law

I referred to Charles Handy earlier (see Introduction). He studied and taught corporate management and developed an interesting analysis of how companies function. With the now well accepted view that companies create their own individuality, he defined four quite distinct cultures that he found in large companies. He gave them names after some of the Greek gods, matching their characteristics. The first he calls the Zeus style company, named after the father of the gods, the paterfamilias, someone who is the initial founder of an entrepreneurial company. The style of leadership is often intuitive, hierarchical, undemocratic and charismatic, with a strong vision and focus.

Ina Meyer Stoll, European co-ordinator for the Global Ecovillage Network, visiting Laird Schaub, veteran community activist, in the tempeh making workshop at Sandhills Community, USA

The second type he called Apollonian, after the god Apollo, who played the lyre and presided over the Oracle at Delphi. This type of company is more organized and subdivided; jobs are defined by roles; the term 'human resources' is used instead of 'labour' or 'manpower;' and individuals can be moved from function to function. This company culture can be described as predictable, bureaucratic, secure, and well documented.

In the third type of company, Handy finds people working in small groups, defining and solving problems and tasks. This he names after Athena, the teacher of spinning and weaving, also associated with wise and good government, and the company is characterized by task-orientated group work. It has a flat horizontal structure; success is very important, while formal qualifications or length of service are much less so. The working groups operate in a matrix, drawing on as many resources as is necessary in order to solve their tasks. In the last type of company Handy found highly qualified individuals doing their own thing, but grouping together for certain mutual benefits. He called this after Dionysus, the god who taught humankind to grow grapes and make wine, and gave examples such as a partnership in a law firm having a shared office and secretarial resources, or a medical practice with several doctors sharing a clinic. This is the other end of the spectrum from the Zeus style company; here there are no bosses. He calls universities typical Dionysian institutions.

This analysis can also be applied to communities. I think we can all find examples which fit. On the one hand we have the community dominated by the strong leader, the guru-led commune, Zeus style. Moving away from this we will find the community that is more democratic, ruled by laws and regulations, maybe many kibbutzim fit into this category, the influence of Athena. Typical co-housing groups may be described as Athenian or Dionysian. It could be a fun exercise to take some of the communities you know well, and see if they fit into these categories defined by Handy. Maybe some communities have an interesting mixture of two or more.

At one point in his book, Handy hints that there may be an evolutionary development from Zeus through Apollo and Athena to Dionysus:

> Once upon a time, the paterfamilias ruled as Zeus over his table. He was master and all knew it. They did as he commanded when he commanded. Then Apollo became the fashion. To each his duties and status. The men made money, women the food, children the beds. Democracy brought Athena into the family. Not duties but tasks, projects and small group activities became the feature. 'Why don't you both ...?' or, 'Shall we ...?' And now Dionysus? A temporary liaison of individuals. If individual interests start to diverge the liaison cannot be enforced. Joint activities (e.g. holidays) cannot be assumed or imposed, only negotiated. Do organizations follow a similar sort of cultural inevitability? If so, where are we on the route?

I have been asking myself a similar question about community for some years now. Can we perceive a life cycle in community, where they go through similar processes as they grow older? This is what I am exploring in this book.

In the summer of 2004, I made brief visits to several communities in America, and found myself trying to apply this pattern to the places I visited. I used a much simpler and more prosaic subdivision than Handy, thinking merely in terms of youth, maturity and old age. What do these stages look like and feel like in the community context? In the first part of this book we looked at some of the characteristics of youth, the early exuberance and the enthusiasm of a community in its start-up phase. Now, what happens when the community begins to mature? Can we perceive the growing pains of puberty, the often painful transition from childhood to the maturity of adulthood? Don't we recognize that in our own lives? Looking back, can we see the transition from a wild teenage phase to getting stuck into the task of getting an education and a job, rearing children and making something of ourselves and our lives? During that summer I visited young communities such as Dancing Rabbit, to recently stable, like Earthaven, and to the more mature example of The Farm, into its second or third generation. I spent several days in the Amana villages, clearly an example of a strong community movement well into old age, perhaps in its last phase.

Dancing Rabbit, a young community

My visit to Dancing Rabbit community (see below) was very brief, with a group of delegates travelling to a conference in Iowa in the summer of 2004. We had a lightning tour of the place, and I was left with a few vivid images of a relatively young and enthusiastic eco-community. The environmental awareness was present in many small signs: the liberal use of mud or adobe in nearly all the constructions; the fireplace in one of the bigger houses, where the flue was conducted through a bench next to the fireplace, giving a warm sitting area in winter, and spreading additional warmth through the main living room; the use of an old and large agricultural silo as a small residence, lining it with mud-rendered strawbales to give insulation; the re-use of building materials gathered from dumps and building sites.

I felt tremendous enthusiasm, and noticed activity going on everywhere. The place looked like a building site, and there was very little landscaping between houses. Paths wound around heaps of stones, earth and mud. Little gardens had been established here and there, risking avalanches of stray building materials. There was a much larger vegetable garden, a collective contribution to the community's food needs. There were people already living in half-built houses, with unrendered inside walls oozing wires, pipes and cables.

The community was very exciting, still vigorously looking for new members. We were asked if there was anyone in our group who would like to move in. It was very tempting to join that enthusiastic pioneering bunch, still working out their communal culture, and still open to new ideas.

Earthaven

I first heard about Earthaven (see above) when I attended my first Permaculture Design Course on The Farm in Tennessee in 1995. Three of the instructors were then designing the place; the site was in the process of being bought, and we were shown maps and designs of what their vision consisted of.

Nearly a decade later I was back in America, and visited Earthaven twice during the summer of 2004. By then the community had gone through its early stages, which were made much easier by the careful Permaculture planning before anyone moved on site.

Earthaven is a powerhouse of alternative American activity. Two influential magazines are edited there, *Communities Magazine* and *The Permaculture Activist.* Several Permaculture designers live there full time, running design consultancy and courses. This puts the place squarely on the map; many people come for courses, many more read about it. Every Saturday there is a guided tour, well attended when I was there. And there's lots to see!

Houses range from modest single-person units tucked away in the woods to a co-housing group with several families living in one several storey building. Self-build with natural or recycled materials is evident everywhere. There are house gardens, and community gardens throughout the wooded valley, in clearings and around the buildings. Good relationships have been built up with neighbours, and there seems to be lots of interaction between the community and its surroundings.

In Chapter 5, I recommended that communities define a membership process as early as possible. This is a way of growing up fast. If it isn't put in place early on, it has to be worked out in some way pretty soon in order to be able to allow more people to join. As new members apply, they will need an explanation of how the community functions, a map of structures and roles. Having to explain itself to these potential new members encourages the community to take a new look at itself. As decision-making becomes more formal, rules will develop. In some communities these end up being written, discussed and rewritten. The rule of law displaces the rule of the strong leader. Apollo takes over from Zeus.

This is all fine theory, but the really interesting question is how it works when we apply it to real life communities. Some years ago I made a study of the organizational structure of the Norwegian Camphill communities. Here was a chance to see if these neat theories have any meaning in reality.

Jøssåsen village was founded in 1973, and had about forty-five residents when I studied it in 2003. My first impression was a small village still dominated by the group of co-workers who had been there for at least two decades, and who had little need of formal structures to keep running the village. There seemed to be a gulf between those in leadership roles and the new younger volunteers. I didn't perceive a middle group of co-workers. Jøssåsen lies at the end of a country road, and feels isolated, deep in the forests. My impression was that the group of veteran co-workers were also isolated, deep in their own long term experience.

I attended the Jøssåsen village meeting while I was there, and was impressed by what a good atmosphere there was. It said a lot about the village. It was clearly a popular meeting, with its own easygoing dynamic. The place felt secure, the structure seemed stable and predictable. In the long run, though, I felt there would be a problem with finding a strong group of co-workers capable of taking the village onwards. I couldn't see a middle generation being groomed for leadership roles, or a clear structure that could be taken over by those who would replace the present pioneers. Here was an interesting mixture, a group of established Dionysian members who ruled the village like a corporate Zeus. Everyone was comfortable, things ran pretty smoothly, but no-one saw the clouds gathering over the horizon.

The Camphill village Hogganvik, on the west coast of Norway, was also founded in 1973, and here I found a similar situation. It was also a small community still dominated by the co-workers who had been there for at least two decades, and had little need of formal structures to keep running the village. It is the most isolated of the Norwegian Camphill villages, being a whole day's journey from any of the others. The older leaders are becoming tired, and again I couldn't perceive a strong new group ready to take over. This came out most clearly to me when I asked people to talk about how the village had changed over the last few years. There had been more enthusiasm before, but now there was not enough energy to prepare a new group of stable co-workers to take over. The small group of people leading the village for so long had not developed a formal structure which was easily accessible to new co-workers. The management style worked well for them, but could easily be perceived as closed to a new arrival.

Today, nearly three years later, these two villages are experiencing the problems of renewal, and are confronted by the question of how to find that group of experienced but younger co-workers who will lead the village into the future. Unlike individual people, who only get older, communities can experience total renewal by having new groups come in and start afresh. This makes it more complicated to attempt this kind of analysis of growth and development, but infinitely more interesting. It's really important to keep in mind that communities are individuals, each one is unique, each one has its own biography. Within the Camphill communities the situation is even more complicated by the fact that about half the population have mental handicaps, usually referred to as 'villagers' or 'companions' while those who think of themselves as 'normal' call themselves co-workers. Perhaps when looking at organizational structures, villagers play a fairly small role in determining them or indeed maintaining them. The structure is set by co-workers talking about co-workers and not so much about villagers. Despite the often said phrase that 'villagers are central,' they might in fact not be so central when it comes to organizational structure. In my experience as a co-worker involved in a wide range of meetings I find that in the minutes of meetings the majority of things discussed do not directly concern 'villagers,' but are more concerned with 'the village.'

The core of the Camphill village

In Camphill villages, the community is focussed around the work with those who need help to get through life. Not that everyone doesn't need help, but some need more than others. One way of measuring the quality of a society is by seeing how their weakest members are treated. By treating them well, we create a higher quality, something which benefits all the members of that society.

It was commented at a recent meeting of the Norwegian Camphill villages that people with handicaps fulfil a function similar to that of the canary the miners used to take down into the mines. If the air became bad in the mine, the canary would fall ill, and the miners would know it was time to leave. So when we see that our friends with handicaps are suffering, it's time to find out what our society is doing wrong, and make changes to get it right.

In my experience of Camphill, it is those with handicaps who carry a lot of the continuity and stability of the community. By centring our lives around their needs, structures are created that give a rhythm to the passing of time. The day has its periods well defined: meals, work, and cultural activities. When we co-workers forget, our friends remind us.

Since the pioneering days of Camphill, when work was focussed on children with handicaps, there has been an element of paternalism within the villages. This may partly have been an inheritance from the surrounding society, and partly a need when building up a new type of society from scratch. As Camphill has matured, some of the more observant co-workers have perceived this, and even as I write, there are fundamental changes going on within the Camphill world to create a society which is more person-centred.

For me, this is just another indication that a healthy community should be in a continuous process of growth and development.

Clearly there would be no village without some form of social organization, on the other hand, it doesn't seem to matter much to the villagers whether it's democratic or not or whatever. They get by with whatever comes. It may be at the end of the day that whatever form the organizational structure happens to take is really not important, because the villagers fulfil such a special role. They are in many places the most stable element within the village. Co-workers come and go, house parents are replaced and turn over every few years. The villagers see people arrive and depart, and take it all in good spirit. I can only marvel at the capacity of these people, who individually would be so vulnerable in any other kind of society, to cope with the number of co-workers who pass through the villages in the course of the years.

It is the co-workers who burn out, crack up, break down, and in the end leave to follow other stars and ambitions. The villagers, those officially designated as 'weak of mind' carry on working the farm, the gardens, the households and the workshops. They form the basis for celebrating festivals and seeing the seasons swing through the year. Their capacity for remembering past co-workers is prestigious. Without them the co-workers would be lost, endlessly arguing about how to organize their community.

In the *Communities Directory* published by the Federation for Intentional Community, there is a classification called 'Forming communities.' This is defined as:

> Any community that labels itself as such, any community with fewer than four adult members, and any group that has not lived together for at least a year. 'Forming' communities are often, but not always, brand new. Some older existing communities go through periods of reorganization or upheaval in which their populations drop to 'forming' levels.

The second point made here is something that can be seen in community. The situation of villages like Jøssåsen and Hogganvik is not necessarily the end of the story. A small but vibrant group of new members can move in and create a rejuvenation of the community. Other circumstances can precipitate this, like the exodus of over a thousand members of The Farm in Tennessee which happened about a decade after it was founded. This is described in closer detail in Chapter 11.

How much a community can keep its identity even when going through fundamental changes is one of the themes I will explore in later chapters. At this stage, the first signs of maturity, we are concerned with self-definition and creating an identity. Ed Schein, writing about corporate identity comments:

> The emphasis in this early stage is on differentiating oneself from the environment and from other organizations. The organization makes its culture explicit, integrates it as much as possible, and teaches it firmly to newcomers.

This would be the ideal, and I hope that communities that get past the initial enthusiasm of pioneering, then settle down to do exactly this. As with so many things, there needs to be a balance between creating a defined identity and getting stuck in a rut. There is always a danger that the rule of law will fossilize the conditions in a community, that people will pay more attention to the written letter than to the spirit of the law. Dangers are there, stray too far one way or the other, and you fall off the path. Friedrich Glasl, writing in *The Enterprise of the Future,* recognizes this:

> I know many communities which have set themselves such high moral objectives and therefore reject any kind of restriction of individual freedom through community regulation. The result is many types of conflict in these communities.

Here in this discussion there are dangers too, the risk of becoming too theoretical, of trying in the end to shoe-horn what I see in community into some schema that I have thought up. That's one of the reasons I have presented a much more simple idea of Youth, Maturity and Old Age. Handy's titles from the Greek Gods and Brink's seven stage developmental theory are very appealing to me, but when I go to see and experience a community, it's not always so clear exactly how each one fits in. Rather than being frustrated at not finding the exact fit to the theory, I have found that it's better to concentrate on what is really happening, to see the fellowship 'warts and all' for what it is.

It's nice to see that there are real communities out there doing it. That though we strive to create heaven on earth, what we actually achieve is just a little better than the average mess. That communities are human too. That the most peaceful looking rivers have treacherous undercurrents.

8. Kick out the Troublemakers!

In the development of the human being, something happens in the years between about the age of fourteen and twenty-one that establishes the person's identity. This period is often regarded as difficult, and is full of searching, questions are asked, thinking really sets in while the person tries to find him or herself. Christianity recognizes the beginning of this period in the ceremony called Confirmation, literally confirming the young person's faith. Judaism does likewise with the Bar or Bat Mitzvah, also reaffirming the commitment of the individual to his or her religion. Now I'm not suggesting that a community needs to wait a decade and a half before establishing its own identity and self-awareness, but I am trying to find that crossover point, when the young and lively group that pioneered the original impulse to set up a new, intentional and radical community, becomes more mature, more sure of itself and its direction.

Searching questions are asked as a community becomes more mature

Ed Schein makes the following comment in *The Corporate Culture Survival Guide:*

> The most salient cultural characteristic of young organizations is that they are the creation of founders and founding families. The personal beliefs, assumptions, and values of the entrepreneur or founder are imposed on the people he or she hires, and — if the organization is successful — they come to be shared, seen as correct, and eventually taken for granted. The shared beliefs, assumptions, and values then function in the organization as the basic glue that holds it together, the major source of the organization's sense of identity, and the major way of defining its distinctive competence. At this stage, culture is the organization's primary asset.

So he calls it 'culture.' I use the word 'identity.' We might bring in other terms, but they all point to a similar stage, the one following that initial exuberant burst of fun, the foundation of a new community. Unfortunately communities can't remain at that stage too long, either infant mortality sets in, and the community disintegrates, or we pass to a mature stage. What is appropriate at an early stage often becomes inappropriate in later life. Don't be too shocked at the idea of communities breaking up at this early stage. Many, many do. Diane Christian has observed that most do. Statistics from small companies give a rule of thumb that at least half of the small businesses that are set up fail in their first year, and that over half of those that survive that first fateful year will fold in their second. Like in those bad old days when basic hygiene had not yet been invented, child mortality was high, and remains high with human groupings. Maybe we haven't yet introduced human group hygiene, and we are still at the stage that medical science was two hundred years ago. Now doesn't that make you think?

Staying with the analogy of the human being, having a mental age of five is really endearing in a five year old, and a really serious challenge for a fifteen year old. As I tried to point out in the previous chapter, small groups at the beginning of their development often don't have formal decision-making structures. In some cases this initial and healthy anarchism can develop into an ideology, thus blocking the group from moving into a mature phase. Cartwheel was a classic case. It never got over that first phase. There were many reasons for that, but the fact remains that Cartwheel was one of the ecovillage attempts that failed in its aim of setting up community.

One of the situations which can really challenge a group is how to deal with a member who is behaving contrary to the accepted norms of the group. How a group tackles the issue of conflict resolution can be a sign of its maturity. How to develop the empathy to deal with conflicts without hurting individuals is a challenge to all of us. Within community, it might be helpful to regard conflicts as being given to us as gifts, they are the questions and the challenges that open up the opportunity to develop ourselves, to rise up out of ourselves and find new depths, new skills and new truths. Without conflicts there will be less development. You should always be suspicious of communities which claim not to have conflicts.

I was told once that at an international communities gathering in the early 1980s someone asked the kibbutz delegates how they dealt with conflicts. One of the kibbutz members said brashly, 'Oh, we don't have conflicts, when we encounter a problem, we set up a committee and they deal with it!' The person who told me the story was a kibbutz member with a lifetime of experience, who had joined his kibbutz in the 1930s. He was embarrassed and hurt by the shallowness of the answer given. Of course kibbutz has conflicts; in fact, one could easily write the history of the kibbutz movement, nearly a century old already, as a series of crisis, each one following the last one. The depth of the kibbutz experience is one of the reasons I have already, and will carry on, telling anecdotes and stories from kibbutz.

Conflicts arise when people try to work together, they are the inevitable result of meeting each other and having slightly different ideas, or different approaches, or worse still, the mixing up of interpersonal chemistry with the issue at hand. Conflicts often arise from mixing up antipathy and rational thought. If I don't like you, I might fall into the trap of disagreeing with whatever you propose, even though I may rationally agree with it. This is perhaps one of the most common causes of conflict, and the only way to combat it is to embark on an inner task of cultivating empathy, and avoiding antipathy and sympathy.

Community begins with two people

Community arises out of the space between people. Even though there are brave souls out there, trying to create an ecovillage or a community on their own, most of their initial energy will go into trying to attract others to join them. Should these remain on their own, unable to recruit additional members, community will not happen, rather the opposite, hermits will emerge.

In the old days, when people got married, we might talk about their relationship as a thing in itself. We might say 'they have a great marriage' or on a sadder note 'their marriage fell apart'. Community is like a marriage in some ways. There is a feeling that you choose a set of partners, in a completely different way than when someone buys a house and moves into a new neighbourhood.

As the group expands, taking in more members, something develops between them, which we call community.

This takes on a life of its own. Speculating on this thing's growth and development is what this book is about. It has no physical existence, yet we feel it when we visit a community, we don't have problems with describing this as something real; A vibrant community, an exciting community, or even a dysfunctional community. Even our most materialist friends can't help admitting that there is something here which we can perceive, feel, enjoy or dislike.

Look at the pictures here; it's between the individuals. You can't see it, but it exists.

It helps to display these qualities graphically:

Empathy (progress)

Antipathy (engagement) Sympathy

Apathy (regression)

Often, our initial response upon meeting someone new is one of sympathy or antipathy. We often call it the chemistry between people. If it is a very strong feeling, then it may be a sign that there is some bond here, some connection which is asking to be developed. An engagement. Being apathetic about other people is really an unhealthy sign. Concern for our fellows is accepted as being a quality which we try to develop within ourselves and clearly encouraged by those great teachers of interpersonal relationships, our major religions. Buddhism teaches compassion towards all things, Christianity tells us to love our neighbour. The opposite of love is not hate, it's apathy, lack of care. That's why I consider apathy to be regressive, it has the quality of removing us from engaged contact with our fellow human beings and from the world.

At least sympathy and antipathy engage us; they commit us to a connection. But for real self-development to take place, we need to develop empathy, the quality of compassion, of being engaged without either liking or disliking the other — feeling with. Small groups of people working together without being dominated by a hierarchy or a pecking order can develop this quality through their handling of conflicts. These situations give us tremendous opportunities for self-development; for learning how to be with other people; for knocking those sharp corners off our square old selves.

One way that a group can decide to work with this is to learn consensus decision-making. Consensus was already mentioned in Chapter 2. Here we go a little deeper.

Consensus grew out of the Quaker meeting process, which is based on the premise that 'there is something of God' in each and every one of us. This has a long history, stretching right back several hundred years into the beginnings of the Pietist movements of the sixteenth and seventeenth centuries. This has special significance for the Camphill villages. Karl König, the founder of the first Camphill village in Scotland, once wrote an essay about the roots of the ideology of the movement, and one of the three people he names is Count Zinzendorf, who was a central figure in the development of the Hutterites, whose later descendants are the Bruderhof. König himself spent some time working in a Moravian community in the 1930s, and his wife Tilla came from this community. The Moravians, Quakers and Hutterites have common ancestry, and Zinzendorf is shared by all of them. During the late 1960s the protest movements in America, the civil rights movements for black equality and anti-war groups working at stopping America's engagement in the Vietnam War, needed good decision-making processes. Many Quakers were involved in these protest movements, and their consensus-type meeting process came to be adapted by these protest groups. Participants saw how well they could work. The Quakers' basic premise became secularized to 'there is something of the truth in every one of us.'

In any discussion where people don't agree, each person has something to contribute to the answer. The challenge is to find a way of giving each person the feeling that even if they don't agree, they have the chance to be heard, and be listened to. If we use the old-fashioned democratic process of dividing up into two or more conflicting blocs, and then taking a vote where the decision rests with the majority, there will nearly always be a dissatisfied minority: those who lose the vote. Black and white decisions rarely satisfy everyone; they create winners and losers, and it seldom feels good to be one of the losers. Doing this continuously will give us significantly large numbers of people in our community who don't feel good, a lot of the time. This doesn't sound like a good situation to me.

Being together in Community

When individuals get together and begin working together on a project, there are a number of different relationships that occur between them. Sometimes we call it interpersonal chemistry, and either bemoan the fact that it just isn't working, or we congratulate ourselves on how well we get on together.

It may be worth looking a little closer at what this means — interpersonal chemistry. If there is no feeling for the other person, a complete lack of interest, we call this apathy, and I hope that this is rare; after all, living in community implies an interest in the other, indeed, demands it. Generally it's more likely that either sympathy or antipathy develops. If a relationship starts off antipathetically, it's off to a bad start, co-operation is going to be difficult, and conflicts can be seen lurking on the horizon. We might want some help from others to work through that, if we have to get along in order for the

task to be carried out. I'm sure we're all familiar with that situation. Sympathy has its dangers too. It can easily slide into a power game, especially if one of the partners is more powerful or capable than the other. Despite the risks of that, I think it's better to start off sympathetically than the opposite. At least it's more pleasant to work with.

The winner is of course empathy, a 'feeling with.' This can be a really good starting point for working together, and has neither the discomfort of antipathy, nor the risks of sympathy. Out of empathy we can build up a real liking for the other person, and much more understanding. Training ourselves to feel what the other person feels, to set ourselves in their place, and to keep our own likes or dislikes of them out of the way, is a best bet attitude to creating living community.

It isn't easy, but there are lots of good tips in the major religions. Even if you don't want all their theological baggage, you might want to consider Buddha's compassion for all creatures, or Jesus' advice that we love our neighbour as ourselves.

In order for consensus to work, Betty Didcoct, writing in Hildur Jackson's book *Creating Harmony,* lists a number of conditions:

— There should be a clear agreement about the purpose of the group
— Each person needs to respect everyone else and trust that each person is doing their best
— There needs to be sufficient time to devote to discussions
— There needs to be a good facilitator, preferably trained and experienced in consensus process

At the same time as consensus is a group-building process, it is also an individual self-development process. As issues come up, each person might be asked to look into him or her self. Seek to understand. Instead of rejecting ideas, ask how the idea that you are not agreeing to might work. Break your emotional ties to your ideas, put them forward, and let them become the property of the group. Be flexible, stay open and let go of petty hurts. These are all virtues that we may aspire to in any phase of our lives. Disagreement and conflict will inevitably arise when groups try to find ways of working together. They should not be avoided or ignored, Betty calls them the 'grist for finding clarity and creative solutions.'

In order not to get totally stuck over issues that are hard to resolve, consensus facilitators have evolved two fallback positions. One is often called 'stepping aside,' and is used when a person can't agree with a decision, but is willing to let the group carry it out. They may be excused from having to participate actively in this carrying out. The second is called 'blocking,' and is more serious; it has also been called the 'veto of one.' Betty Didcoct comments that in her twenty-five years of facilitating consensus she has only seen it used twice, and both times it proved to be wise. If several people block a decision, it is a clear indication that either not enough work has been done, or that this is indeed not a good decision. If one person repeatedly blocks the decisions of a group, it may be legitimate to ask whether that individual really shares the basic values and goals of the group. She goes on: 'The consensus process is more than just a decision-making technique. It is a way of being, a way of listening, and a way of understanding.'

A favourite topic among communities is the discussion of what to do with people who break the rules. This becomes even more interesting when the rules become hardened, but the wish to become more flexible still predominates. In one well established income-sharing community that I heard about, a key member with several decades of experience was found to have bought a part-share in an aeroplane, which had cost quite a considerable sum. This was not declared to the other members, but was discovered when the auditor found a large sum unaccounted for. It provoked a major crisis within the community, the end result being that the aeroplane-buying member was asked to take a free year in order to think through his connection to the rest of the group. To some members this was a clear-cut case of someone overstepping the bounds of trust, and expulsion was deemed to be a fair ending. Others thought that this was grossly unfair to someone who had been a contributing member for so long. This was a classic case of that old conflict between wanting to keep to the rules and wanting to retain flexibility.

At first sight it may seem sensible to try and avoid conflicts. If someone could just tell how not to have arguments! Arnold Mindell, in his *Sitting In the Fire,* has a slightly different take on this view, and comments:

> Organizations and communities do not fail because of their problems, nor do they necessarily succeed because they solve them. Problems will always exist. Communities that succeed open up to the unknown during periods of crisis.
>
> We come together by following the flow — through operating as a community, having trouble, almost going to pieces and then coming together.
>
> Once a community gets together in an open forum and deals with its most difficult issues, it knows itself from a new angle. The atmosphere improves.
>
> You find that a battle does not mean the end of the world, but the beginning of the river called community.

Conflicts may be the challenges that are sent us to give us the opportunity to grow and develop.

Using the circle to create community

The micro-geography of our meeting culture is very important. When I was active in the trades union movement in England, the usual pattern was to sit in rows. The main characters sat facing the assembly, with a table in front of them. It felt very comfortable and familiar. This was how classrooms were arranged when I grew up.

The implications of this layout were that most of those present had very little to say to each other, but had to address the 'leaders' who possessed higher status, with their table in front of them, upon which they could put their important papers, and make notes. They might also have glasses of water, or ash-trays. Yes, folks, in those days we could smoke as much as liked at meetings! Notice how much our attitudes change. Today, it would be unthinkable to smoke at a meeting; thirty years ago it would have been unthinkable not to.

When we arrange ourselves in a (smoke-free) circle, we create a different set of relationships. We are all more or less equal, we all have something to say directly to each other,

and we can see and hear the speaker. The leader may now be called the focalizer, or the mediator, or possibly the convenor. Specified roles, such as note-taker or heart-keeper can easily be rotated or passed from one to another without disturbing the seating arrangement. This implies we all have something to contribute to the group, we all have something we can learn from each other, and that leaders are servants, helping the group reach their destination.

Our cultures change, and cultural references become recognizable. Sometimes, these layouts are fossilized. This is often the case at universities or other educational establishments which were built in the heyday of the 'sitting in rows with the leader at the front' era. The lecture hall is often really advanced, with all kinds of gadgets which make presentations easy, but the seats are bolted to the floor. In rows! I've tried all kinds of tricks, but it's not easy. At one seminar, in Turkey, the weather was good enough for us to move out onto the grass outside, and abandon the lecture hall altogether.

9. Building the Building Business

Clem Gorman spoke prophetically in 1975 when he wrote in *People Together:*

> Perhaps the day will come when premises can be custom-built for communal projects by ecologically-conscious architects, but for the present all communal projects live in places built for other purposes, and converted.

The day has indeed come; in fact it's been here for quite a while. Even here in Norway, which is not exactly one of the most active ecovillage countries, we have several architects and planners who have specialized in building ecovillages. One of them has written a book about it. Unfortunately, it's in Norwegian, so inaccessible to non-Norwegian readers, but for the lucky few, check out Rolf Jacobsen's book *Tun, bygninger og Økologi* (Courtyards, buildings and ecology).

Experience in converting older buildings to community use is now very widespread. A community which builds its own houses, or retrofits existing ones, will soon gather a wide range of skills which can be marketed in the outside world. This can create additional income, and may help to spread new techniques throughout a wider society. The ecovillage at Lebensgarten has developed at least two building and consultancy businesses founded upon these amassed skills. This is a real win-win situation. The ecovillage gets built, and both the practice of ecological building and the idea of it spread into the surrounding society. Members of the ecovillage are gainfully employed, both inside the community, and outside, bringing in additional income. Inevitably, when you get involved in building, you create contacts with suppliers of materials and with other builders and trades people, electricians, plumbers, diggers and carpenters. This opens up many opportunities for creating co-operation, for getting to know neighbours and others in the locality, and for getting cheap deals in materials. Given that you are doing a good day's work for a good day's pay, and that you are careful to carry

Hurdal ecovillage at the beginning

When the first group of pioneers moved into the ecovillage at Hurdal, their idea was to build a cluster of temporary homes as quickly as possible. They knew it would take several years to establish their legal status, work out their zoning plan and get it accepted by the local authorities, and begin building their final houses. Along the way they had to work out finances, attract more members, and survive for long enough to get all this done.

There was one family-sized house on the site, which was crammed to capacity, and also served as the main kitchen and meeting room. Trailers were brought in to live in, and during the summer, some people lived in teepees and yurts. Summers are fine in Hurdal, but winters are a challenge, sometimes up to a metre of snow, and temperatures dropping to below –20 degrees centigrade. So their first task was to build quickly, simply and cheaply, homes they could move into soon. Together with their architect, Rolf Jacobsen, they designed strawbale houses built on a gravel plinth, using euro-pallets for flooring support. In addition, a log cabin and a prefabricated house were built, and an agreement with the importer of a new experimental housing system. Unfortunately this importer folded after a year, and the idea of Hurdal being the show site and agency for these buildings went the same way. But the first two of these new houses are there, and by all accounts, doing a good job.

Many practical building skills were learnt during this process. Lots of mistakes were made and corrected, and many people came through, attending workshops, work-ins and seminars. Local builders and engineers also got involved. Building the first houses at Hurdal connected the ecovillage group to the local economy and the wider society.

GEN meeting at Hurdal, with completed strawbale buildings in the background

out quality work, you can establish a good reputation and good relationships with your neighbours. Actions speak louder than words, and even if you might look freaky, or have weird ideas, if you are seen to be good workers, much will be forgiven.

Stephen Gaskin of The Farm in Tennessee maintains that one of the reasons that they were eventually accepted so well in the redneck backwoods where they settled in 1971 was that they were prepared to help their neighbours do any job that needed doing, doing it well and quickly, and not being over concerned with squeezing the best financial deal out of any business transaction. In remote backwoods places, where people know each other, and have lived around each other for generations, your reputation often counts for more than making a quick buck and moving on. I was once working on farms in a remote part of mountain Norway, and asked a friend how much I should charge per hour. He advised me not to set a figure, but to work really hard and well, and then ask how much the farmer was willing to pay afterwards, making sure there were other people present. He explained that if I was seen to be a good and worthy worker, it would reflect badly upon the farmer if he underpaid me, but raise his status if he was seen to be generous towards someone who was worth a good wage.

All this sounds very hunky-dory, but there are risks involved in setting up a small business which has to compete, which sets as one of its aims the generation of profit, and which may create an entrepreneurial class within the community. This is not to suggest that a community should not create businesses which trade with the outside world. Not at all, I think it's one of the more important activities that a community can embark upon. My concern here is to point out some of the possible pitfalls which inevitably open up before the budding community entrepreneur.

The relationship between those who go out to work and those who stay at home needs to be dealt with sensitively. It can be very exciting to set off in the pick-up in the morning, do interesting things in the neighbourhood and meet lots of other interesting folks, and come back in the evening counting good cash for work done. What do the stay-at-homes do? Look after the children, weed the vegetable garden, wash clothes and tidy up. And they didn't even earn any money doing these things! Or it could be the reverse. Back in the community there's organic mulch gardening going on, the Permaculture herb spiral is getting its final touches, a dung and clay floor is being polished with a blend of natural oils and there's a workshop for the kids on music and puppet theatre. The folks going out to work set off early, taking sandwiches and a thermos of tea, they work hard all day, come back exhausted, and are too tired to take part in the talk on Permaculture and peak oil scheduled for the evening.

In both these imagined and exaggerated scenarios a great deal of sensitivity would be required to avoid conflicts developing between those who stay at home and those who go out.

Individuals who bring an income into an income-sharing community are high-risk potential conflict generators. If I put a great deal of work into a project, for example, writing a book about ecovillages, I will have a special relationship to the income that this generates from the publisher. Some of it I will need for expenses, buying certain books, subscribing to magazines and newsletters, perhaps travelling to meetings and conferences. Cultivating the network that is needed in order to keep in touch with what's going on in the ecovillage world. It's easy to get the feel that it's my money. I generated it, I need it. Other people in the income-sharing group see me zipping off to interesting places and meeting interesting people, and resent it. The tension between those who stay at home and those who travel around doing other stuff can be very hard to bear. The same tension arises in community movements between those who work for the movement and those who don't. As the community becomes more mature, and develops more complex structures, there will be an increasing need for administrators and negotiators, and these are usually specialized tasks that are difficult to rotate. They also have an air of glamour or prestige attached to them. The stability of maturity brings with it increasingly complex conflict situations.

There's a saying that it's the shoemaker's son who goes barefoot. At The Farm in Tennessee, if you want to see the result of their building work, you will have to travel quite far away. When they bought the place in 1971, they had buses, trailers and tents, and lived simply, cheaply and very crowded for a long time. Their houses were built from found, scrounged or cheaply bought scrap materials, whatever they could lay their hands on. Often they just added to whatever was there, building around what they had started with. I was told that if you scratch the wall of The Ecovillage Training Centre you will find some of the original ex-army supply tent that was the original living quarter on that particular site. Check it out on your next visit; this is the new science of ecovillage archaeology! If you want to see the results of the work of The Farm building crew, if you want to see Farm architecture, you have to go Central America, Guatemala and Nicaragua. Or to Botswana. Through setting up the aid agency 'Plenty,' building work has been carried out in all these places, often involving sending fleets of buses and semi-trailers down into Central America in the wake of earthquakes and other natural disasters. This began modestly in 1974 with the idea of helping communities outside The Farm. A year later, Plenty began co-operating with the Mennonite Central Committee and sent help to Haiti which was suffering from famine, and to Honduras after a hurricane disaster. Things really began to move in 1976 when an earthquake struck Guatemala, killing over twenty thousand people. Fleets of semi-trailers drove to Central America to build houses and develop aid schemes.

Reminders of youth

When The Farm was established in 1971, the people were all living in converted buses, trailers or other mobile homes. There was nowhere else to live, so they stayed in these homes while building more permanent houses out of found, given or secondhand materials. The houses were sometimes built around existing buses, trailers or tents. In this picture you can see the improved door, and an attempt to pop up a higher ceiling, on an old school bus.

Many stories abound about the old pioneering days, when conditions were crowded and primitive, money was scarce and, at times, so was food. One of the ground rules of the community was that everyone shared, and the only acceptable income was earned through work. No living off benefits or social security. One particular bad winter occurred in those early years, and virtually the only food left was wheat. You still meet people at The Farm today who talk about 'wheat berry winter.' These are some of the myths of the early days which help to forge the identity and culture of a community.

These old buses, lying around in the forest are part of the history of The Farm. Someone commented that they show the worst side of the throw-away, polluting culture of mainstream North America, but I had a different take. I saw these buses as living history.

The full range of architecture of The Farm Building Crew, however, is to be found in far-flung corners of the earth, in Gautemala, Mexico, Botswana. The houses they built for themselves in Tennessee were scrappy and improvised compared to the building they did for disasters abroad. Through the aid agency 'Plenty,' set up by Farm people, they travelled all over the world, helping to stave off hunger and homelessness.

Helping to build a village in Russia

Volunteer building trip to Russia. Courtesy of Rolf Jacobsen, Norway

The Bridge Building School (Brobyggerskolen) was originally intended to train young people in ecological building techniques, working on specific projects in Eastern Europe. This was what had attracted me to move to Norway in 2000. During that first year, we set up a stand at the highly popular Alternative Fair in Oslo, trying to recruit more students.

Today, six years later, the idea has metamorphosed through various changes, and now consists of two projects. One is regular trips to Eastern Europe with a group of volunteer builders, architects, and others. Funding comes from a number of sources, and not only is this a helpful project for the communities in Eastern Europe that need more and better housing, but it's also a way of training people there in ecological building techniques.

The other project is a building group for the Norwegian Camphill movement, focussing on accredited apprenticeship teaching of ecological building and renovation of older buildings. This is still being discussed at national level, but has already begun in a limited way in two Camphill communities here in Norway, with two apprentices.

The Camphill movement has had its own architecture office for years, and has even developed its own style of building, best documented by Joan Allen in *Living Buildings,* mentioned in Chapter 6. The actual construction work was usually carried out by local builders, usually working with a co-worker from the village. For many years Lars Henrik Nesheim was the Norwegian co-worker co-ordinating such building activity. He had a background in building engineering, and built up a really wide network of contacts in the anthroposophical movement, in education, and in ecological building. He had vision and charisma. In the mid-1990s, he moved to Svetlana Village in Russia, a community which had been nurtured by the Norwegian Camphill movement, and began construction of a strawbale storage building. Returning to Norway, he set up the Bridge Building School (Brobyggerskolen), which attracted me to move from kibbutz in Israel to Camphill in Norway.

There were many bridges to be built. The first and most obvious was a bridge between Norway and Russia, continuing the relationship between the relatively rich Norwegian Camphill villages and the much poorer villages being set up in the former Soviet Union, Estonia, Latvia and Russia. Another bridge was between building and ecology. There was then, and still is today, a pressing need for the building world to adopt more environmentally friendly technologies and practices. In a more esoteric vein, we wanted to build a bridge between the physical and the spiritual, between the idea of design and the concrete realization of such ideas.

Lars Henrik drew up a course lasting for five months, dominated by a lengthy stay in Eastern Europe doing real construction work, sandwiched between two shorter stays in Norway where we covered theory, a wide-ranging education, from building science statics to the history of philosophy. We ran this twice, the first time in 2000, and again in the summer of 2001. We discovered early on that without some kind of recognition or backing, it was very difficult to attract enough students, or make the programme financially viable. By 2005 the Bridge Building School was officially wound up, but during those intervening years it had spawned a number of other initiatives, some of which are up and running. The building in which the original course was held is now an independent seminar and conference centre, hosting a wide range of groups from within the Camphill movement, and also outside. An ever-expanding variety of educational courses are being offered to those with mental handicaps. Groups of Camphill builders are going to Russia two or three times a year to carry on teaching environmental building techniques, and there are plans afoot to create an ecological building apprentice programme within the Camphill village movement in Norway.

This is another example of how the building sphere is creating an interface between the ecologically conscious community and the outside world. If one of the leading aims and characteristics of an ecological community is its aim of social improvement, then one measure of its success must be how much it interacts with that surrounding society. Once a community attains a certain degree of maturity, it must develop a clear identity, and be able to communicate that to the outside world. You don't have to be a community member to see the pattern here. Ed Schein, writing about corporate culture, sees the same thing in companies, and recognizes how valuable this is:

> Culture is the shared tacit assumptions of a group that it has learned in coping with external tasks and dealing with internal relationships. Although culture manifests itself in overt behaviour, rituals, artefacts, climate, and espoused values, its essence is the shared tacit assumptions.
>
> Culture is, therefore, the product of social learning. Ways of thinking and behaviour that are shared and that work become elements of the culture.
>
> A given organization's culture is 'right' so long as the organization succeeds in its primary task.
>
> Culture is a group phenomenon. It is shared tacit assumptions.
>
> Always think initially of the culture as your source of strength. It is the residue of your past successes.

10. Selling Yourself in the Wider Economy

The Farm in Tennessee is big; it occupies an area of about three square miles, most of it flattish, a kind of watershed, with forested valleys falling off to most sides. Walking along the roads that connect the various housing clusters you get an impression of rural peace; there is very little traffic, there are few cars and only occasionally someone passes you riding a bicycle. People seem friendly; everyone waves. Everyone looks like someone you know, or would like to get to know. The big fields that were so intensively cultivated in the 1970s are now mostly grazing, fenced in, with trees framing them in the distance where they dip into the valleys. There are horses grazing. During the first few years of The Farm's existence, the community experienced tremendous growth, reaching nearly 1,500 people in the early 1980s.

The Farm

They practised a collective economy, but income could not keep up with expenses, and in 1983 The Farm decollectivized, and soon the population dropped to a few hundred. There were not enough hands to carry out the intensive cultivation of the first years, and the large fields went over to pasture.

I set off in showers of light rain to visit the Mushroompeople workshop. Most of the houses, workshops, farm buildings and offices are set off into the woods. There are a few dilapidated buildings in need of repair, there is still a slight feel of desolation, of loneliness even. The woods seem dark, dripping, damp. At least it's not cold; actually the opposite, clammy, moist, a little oppressive. The smell of earth and leaf mould.

When I get there I find a long low building. There are stacks of mildewing timbers in the forest around. Here is where they grow mushrooms, specifically shitake. How this came about is a lesson in ecological economics. At a time when The Farm was hard up for cash, one suggestion was to clear cut the forests for desperately needed money. Economically it might have made sense; it would turn some of the natural resources into capital to get the place moving. But of course there were gut feelings against cutting down trees, altering the ecology, destroying the ecosystem. Somehow the connection was made between the expanding market for shitake mushrooms in the US and the conditions for growing them here at The Farm; these conditions were exactly what they possessed in their mature forests. Shitake grows on fallen logs in broadleaf forests with warm moist summers and cold but not frozen winters. By growing shitake they could both keep the forests and make money. Shitake mushrooms can be dried, and sent by post, their value is so high and weight so low when dried, that the postal expenses are low. This makes marketing and distribution relatively easy for a place in such a remote area as The Farm. The clincher was that you could make much more money per acre growing shitake than the value of the standing timber. Every year! Clearly, it would make much more sense economically to grow shitake than to clear cut the forest, and you get to keep the forest ecology as a free side effect. Everyone wins.

So The Farm began growing shitake, and found the market to be expanding fast. Shitake is tasty, healthy, easy to prepare, and very sought after. So much so that others wanted to grow shitake too, and they turned to The Farm for advice. A side business was established helping people grow shitake; a magazine was launched, seeding spores sold, consultancy given, and these activities were found to be as profitable as growing the shitake itself. Nothing succeeds like success.

When I walked in I found a bookshop, a resource centre, a row of computers, as well as the equipment needed to dry, sort, pack and send the finished products. It was a quiet Saturday morning, Frank Michael was alone minding the store, and we talked about business, the weather, and the Permaculture course that had just finished. It was as if I had known him for years. We sat and talked, looked at books, and at designs for greenhouses that he had made.

Mushrooms feed on dead organic matter, breaking it further down in its metamorphoses back to earth. Mushroompeople is a business built on decay. How did Frank feel about that? He said: 'All plants live off decaying organic matter in the soil, turning it into nutrients for themselves, and in turn some of these plants are nutritious for us. Some mushrooms are poisonous, but so are some plants. Actually, mushrooms are very nutritious, especially shitake which has healing properties as well as just food value. Shitake will enhance your health.'

Growing and selling food can be carried out in many ways. It can be market gardening of vegetables, it could be a classic biodynamic farm with a mixed production of animals and crops, or it can be, as The Farm's Mushroompeople show, a niche product which fulfils a number of functions. In every case, the distribution of the food products will bring you into contact with a wider group of people. Unlike a building business, which sells skilled labour, in this case we are selling a product.

Selling this food brings the community, or at least part of it, into close connection with people outside the community. How far can the alternative ideals of the community be carried into the outside world, and how deeply do the underlying principles of the wider society penetrate into the community? Ideals can influence in both directions. Straight selling, either direct through a farm shop, or wholesale through a distributor, will bring you into the world of conventional economics: buying and selling, confronting market prices, and distribution percentages. This world has its own fascination, and as long as you keep good financial records, after a few rounds you should find how to stabilize and cover your production expenses.

Profitable fungi

During the early days at The Farm, money was in short supply, and various ideas were floated with the aim of getting more cash in hand. One idea was to clear cut the forest they had bought, and collect the cash for the timber. This was dismissed out of hand as being anti-ecological, but possibly made some of them look closer at the forest and what it could be used for.

To cut a long story short, someone discovered that the dead and rotting trees in broadleaf forest surrounding The Farm were the ideal environment to grow shitake mushrooms. So began the business 'Mushroompeople.' Shitake mushrooms are highly nutritious, and have certain important health-giving properties. They can easily be dried, and are easy to ship using the postal system; high value, low bulk, low weight. As the business grew, so did the number of requests from people all over the United States for advice on growing the mushrooms at home. This created a whole new sideline of selling spores and providing support to home growers.

Today 'Mushroompeople' is a textbook example of a business based on the local ecology, which maintains that local ecology, and uses the inbuilt characteristics of the product to create a nationwide network.

We could at a pinch sleep outside under the stars, and if the weather is warm we could do without clothes, but we won't live very long if we don't eat. In one way or another we are all farmers. Either we grow the food ourselves, or we pay someone else to do the growing. Somehow, in the western industrialized world, we are paying our farmers less and less, either subsidizing them through a taxation system, or simply driving them out of business by buying cheaply from other countries (where they pay the farmers pitifully and shamefully low prices) or buying from dreadfully polluting agribusiness industries. Whatever we do, we do it mostly through the big supermarket chains, and the food is shipped by road, rail, air and sea for thousands of kilometres round the world. (Unfortunately, less and less is shipped by rail these days.) As the organic food market expands, it is also being co-opted by the supermarkets. Why should I feel a glow of satisfaction when I come across major conventional supermarkets with whole shelves devoted to organic produce? What was that dream? That all farming would be organic, and that the supermarkets would sell only organic? Now I realize that this would only put the organic farmer where the industrial chemical farmer is today, at the mercy of price fluctuations, distribution percentages, vast food miles, packaging and the resulting loss of fresh quality. The farmer would not know who will eat his produce, and the family sitting down to a good organic meal would not know where their food is grown.

There *are* other ways of supporting the farmer, even better ways of ensuring that he or she has a secure and reasonable income, freeing the farmer from paralysing economic fears, and allowing him to concentrate on growing good food for people that he feels connected to. I first came across this system in my initial courses in Permaculture, and it was only later, when I became more involved in the world of Camphill, that I learnt what a central role biodynamic farming and Camphill communities have played in the development of Community Supported Agriculture (CSA).

First a brief explanation of what a classic CSA is like. Imagine a club with two types of membership. The first is composed of people, families, who agree to pay a certain amount each week or month for an agreed period, it could be a growing season, or a whole calendar year. In return they will get a box of vegetables on a regular basis, weekly or monthly, containing whatever is ripe at that time. The other members are the gardeners, who receive a regular payment, and spend their time gardening and delivering or making available the boxes of produce. Most CSA's need a minimum of twenty to thirty families to be viable, and rarely grow too large. Most people within the CSA will know each other.

The classic CSA story seems to have originated in Germany in the 1980s, and moved to the USA during the 1990s, experiencing a rapid growth and distribution, so much so that a national advisory organization was set up in the US to help new ones be established. People researching CSA's cannot find one single example of chemical/industrial farming using the CSA model for marketing. There is also a much higher percentage of biodynamic farms using CSA than the percentage of biodynamic farms in general. Camphill communities also seemed to have played a central role in developing CSA's, especially in North America. This is actually not so surprising, as the common income-sharing type economy practised by Camphill already ensures the farmer and gardener a reasonable and secure financial situation, and the products are often distributed free to the households within the village. It's fair to say that Camphill was practising CSA decades before the concept was even thought up. You can read more about the CSA story in the excellent book by Trauger Groh and Steven McFadden.

In short, in a CSA situation, you pay for production rather than the produce.

Because it takes a couple of years to set up a CSA and really get it functioning, you need to have an established and mature community behind you if what you want to set up is to distribute the products of the farm or garden. A successfully functioning CSA would be a symptom of a mature community. At the same time it is a perfect way to distribute the ideals of an ecovillage. Organic or biodynamic food production, financial security for everyone doing useful work, personal, face to face contact between people who are economically tied together. Most CSA's experience that there are side effects that come as a result of their CSA. Newsletters help to keep the club going, they contain recipes for new or unusual crops, the odd advertisement for needs or services, invitations to parties and celebrations. Many CSA's celebrate together, harvest festivals or regular seasonal festivals. If the CSA is based in a community, it would be natural to include the external CSA members in the community celebrations, and this would be an invaluable interface with the wider society in the locality.

Nancy Foy studied the management of IBM, and teaches management theory to big business. In her studies of how groups work together, she also sees how these processes develop. In *The Yin and Yang of Organizations,* she writes:

> The welding together of a working group, temporary or permanent, is a curious process, quite similar to individual growth. It involves growing confidence, expanding horizons, shared achievement, and enough time spent together in both formal and informal activities to build trust, shared confidence, and a sense of how the various members fit together.

The development of a CSA would give a mature community a tremendous opportunity to establish a close working relationship with a wider group of individuals, not just supplying them with organic food, but in addition allowing their ecological ideals to seep into the surrounding society. Once you have the idea of establishing an alternative economic relationship, there is no reason why this can't be expanded into other areas. Very recently I read in an American anthroposophical magazine, *Lilipoh,* of a health clinic operating in a similar way. The article was called 'Community Supported Anthroposophic Medicine,' and the economic set-up was very similar to a classic CSA. Each family committed itself to a monthly payment, and availed itself of the services of the clinic as required. The professional health workers were assured of a regular income, and the member families of health care when they needed it.

In 1977, in a completely desert location in Egypt, Ibrahim Abouleish set up the Sekem farm and community, inspired by a mixture of Rudolf Steiner's anthroposophy and Islam, a movement which is now creating real and big change in that country.

A school within the village

When Solborg Camphill village was founded, in 1977, it was based in buildings which had been a school for the deaf. Originally a traditional farm, the main building dates from the 1760s.

As the co-workers grew and multiplied, their need for a kindergarten became more acute, and at first a room was set aside for this. The years went by, the children grew and became more numerous, the room set aside became rooms, a teacher was hired, and the Steiner School became a fact. Children joined from the outside, and the school began to take on a life of its own.

When we joined Solborg in 2000, negotiations had already been underway between the village and the school, which had for years been its own legal entity, a recognized Steiner School within the Norwegian Steiner School Federation. These negotiations were based on the fact that without owning the buildings, the school could not raise a loan to carry out the repairs and maintenance, but more importantly, build the extension they needed to house the top three classes. These talks carried on for another couple of years, resulting in the school leasing the land, and buying the buildings for a nominal sum. The decks were cleared, a loan was raised, and as with most Steiner Schools, the parents played a big part in planning and carrying out the work.

The school has become a meeting place between village people and the wider society. Out of the nearly one hundred children in the kindergarten and school, ten are from our village. The other 90% give us a great opportunity to meet others who live in our locality, and all the various school events are times when we work together and celebrate together.

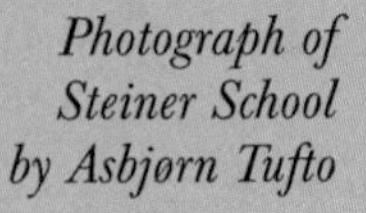

Photograph of Steiner School by Asbjørn Tufto

Steiner spoke at length about the idea of associative economics, the setting up of an alternative to the capitalist profit-oriented company. In his autobiographical account, *Sekem,* Abouleish wrote:

> I imagine an association where the whole value-adding chain becomes transparent. It has to start with the consumer. He is asked which product he wants, what its quality should be, and how much he is willing to pay for it. The distributors agree on a percentage of this known final price for themselves. Finally the producer is given a price, and he knows the conditions which have made it. Everyone involved in the chain is under the obligation to keep to the price they have agreed on, and to deliver the product to the consumer in the state he desired it. An association is founded on an agreement that gives security to all involved. The basis of an association is thus mutual trust or, in other words: economy based on fraternity. Everyone involved in the economic process is aware of the others and recognizes their mutual dependence.

Today the Sekem organization is a charitable trust based on the Sekem community, but in partnership with over a thousand farms throughout the country, employing over ten thousand people. Sekem is a living example of how organic production, combined with associative economics, can effect massive change within a whole country.

We see here how the development of specific skills such as ecological building or organic farming can interface with the outside society in such a way as to create a supporting economy for the community and at the same time spreading the new technology. The same idea can be applied to virtually any technology that the community develops: waste water treatment, non-polluting energy sources, conflict resolution techniques or educational programmes.

Camphill Solborg, where I now live, was founded twenty-nine years ago, and within a couple of years there were enough small kids amongst the co-workers to warrant the setting up of a kindergarten run on Waldorf educational lines. To begin with, this was run by the co-workers for themselves, but within a few years an outside teacher was hired, a few co-worker families had left the village but remained within the locality, outside kids joined the budding school, and there it was, an independent Steiner school within the village. Eventually it established its own status as a charitable trust, affiliated with the national organization of Steiner schools, and got its own management group, board of trustees and statutes. When I joined Solborg in 2000 negotiations were under way for the school to buy the buildings it was using, leasing the land on a long term and secure leasehold. Today the school is a completely independent legal entity, owning its own buildings, with twice as many people attending as there are in the Camphill village which it sits in the centre of. Ninety percent of the pupils come from the surrounding area. There are numerous benefits for both school and village that the two are in such close proximity. Several co-worker families with children were attracted here because of the school being within safe walking distance of their home and workplace. The village shop sells significantly to teacher and parent families as well as to the village. It's only a matter of time before a CSA is established, based on the biodynamic gardens of the village, with a core group of potential customers within the teacher and parent families.

Some communities are inspired by ideals of a simpler life. As technology becomes more complex, the community may move in a direction not originally foreseen. When you come to examine such a worn cliché as 'the simple life' you will find a great number of complex skills. You try setting up a loom to weave the cloth that you then need to cut and sew into clothes. Not exactly simple! Whatever stage of technology you aspire to, you will soon learn skills that have a degree of specialization that qualify them for marketability. These skills, developed by the specialists in the community, may have high earning power, either in the wider society, or for other groups in early stages of development. You can become a consultant. You can teach courses and give advice. You could become an expert.

What we need is a balance, and that might be one of the characteristics that we are looking for in the mature community. A balance between the risks of going out to meet the wider society, and the returns that such a meeting will bring in terms of greater economic stability, and the dissemination of the ecovillage ideals.

We have already mentioned Nancy Foy more than once. What caught my eye in her book *The Yin and Yang of Organizations* was her search for patterns, and her use of the ancient Chinese yin and yang theory of balance, applying it to groups and organizations:

> The Chinese believed that everything in the universe depends on these two processes (yin and yang) being in balance, though it is a constantly moving, changing, dynamic balance. It is a useful picture for organizations; if the conceptual, intellectual yangs ran everything, things would gradually get out of balance, and analysis and learned papers would be its output. If the realistic, survival-oriented, pragmatic yins ran everything, it would also be unbalanced, and we would live in a world of ball-bearing factories. We need both, in moderate balance. I believe that these two hemispheres, this yin and yang nature, also exists in groups and in organizations.

The development of skills and the marketing of them to a wider society is the very core of the ecovillage ideal. We can, should, and are, creating centres for learning. The Farm, as one of the founders of GEN, set up the Ecovillage Training Centre to do exactly that. Lebensgarten, another GEN founder, has several specialized companies, Findhorn is teaching ecovillage skills up to university level, as is Sekem in Egypt. A number of other communities are teaching at university level through the courses arranged by Living Routes. This is how ecovillages are going to make an impact upon society at large.

As the skills needed in a post-oil and climatically disturbed future are developed within communities, they can be disseminated through educational programmes, consultancy and sales. I hope that we can make this more attractive and more realistic than the future offered to us by big business-as-usual: increased armed conflict, increased pollution and sharper internal social alienation.

11. Integrating with the Outside

I have mentioned The Farm in Tennessee several times already. It has an interesting history which has become one of the legends of alternative America. It focuses around the personality of Stephen Gaskin, ex-marine and in the late 1960s teaching modern philosophy at the University of California in San Francisco. He taught an open class, which attracted growing numbers of students. Stephen has a special talent for talking to people, groups of people, large groups of people. He once commented that he had spoken to groups of more than a thousand people more than a thousand times!

The open class grew and he moved into a church hall to accommodate it. This also turned out to be too small, and they found a rock concert hall which was free on Monday nights. Stephen moved in, even more students were attracted, and Monday Night Class was born, a freewheeling debate on modern philosophy, mysticism, spirituality and community. This was San Francisco in the late 1960s, hippies, marijuana, LSD, flower power, an explosion of dynamic ideas, a fusion between east and west, between spirit and matter.

Stephen's fame grew, people heard about him throughout the US, and he began to get invitations to speak and teach in other places. One Monday night he announced he was going on a speaking tour. For him this was no big deal, he lived in a converted school bus, all he needed to do was to drive it out of town and down the highway. But what were the students to do? They wanted to carry on with Monday Night Class.

'Buy buses and join me!' said Stephen.

And that's what they did. In the autumn of 1969 about ninety buses and other living-in vehicles pulled out of San Francisco and set off on a tour of the United States. The trip was to last over a year, and as more people joined on the way they got up to 250 vehicles. People married, babies were born on the road, a couple of people died. They became a community, they wanted to settle down, and the cheapest land they found was in western Tennessee, and so that was how they came to Summertown, setting up The Farm in 1971.

At first they lived in buses and army tents, gradually building houses next to or around these. Even today there are still a few of the old buses out there in the woods, gradually rusting and rotting into the forest. I already mentioned that if you scratch some of the walls, you'll find some of those old tents. Hippie archaeology.

The Farm grew fast throughout the 1970s after settling the land, reaching well over a thousand residents. The economy had never been great, but it grew worse. The initial 'trust-everyone, share-everything' communalism could not go on, it crashed, and the place was totally reorganized pretty much overnight. The community decollectivized, every resident was charged a ground rent, and had to create an income in order to pay the bills. The population dropped to a couple of hundred. But life went on, food was grown, much less now with fewer mouths to feed, and businesses established on a sound economic base. Individuals took personal responsibility, they hired and fired each other, money was made, livings were earned.

The Farm is an example of a community that has managed to survive a life-threatening crisis, managed to take on board a major change, even a renewal. To be sure, it's not utopia, if by that we mean the perfect place. But the real meaning of *utopia* is actually 'no place,' and Albert Bates, the director of The Farm Ecovillage Training Centre, once pointed out that there is another utopia, spelt *eutopia,* which means 'good place.' Eutopia The Farm certainly is!

Was this change from a collective economy to a private one a renewal or a defeat? Was this the inevitable withering away of the ideals of collectivism, allowing that old selfish egotism, materialism and the cult of the ego to take its place? I'm sure that there are as many reactions to this question as those asked. For the residents of The Farm, this was a major experience, a watershed in their community biography, and it deeply affected most of them.

Measuring the success of a community is a very tricky thing. Is it merely a measure of how long the community survives? If this is to be our yardstick and as all communities seem to come to an end sometime, it seems they all fail. This is clearly nonsense, there has to be some other way of measuring success or failure. I think it lies in a subtle sense of their influence. We are breaking new ground here, this is not the hard-fact science of the materialistic age. We are dealing with social factors, with relationships between people. We need intuition, we need opinions, we need to bring in the whole human being, not just the white-coated scientist calculating numbers.

About ten years ago I was invited to The Farm to participate in a Basic Design Course in Permaculture. I was then a member of Kibbutz Gezer in Israel, and this invitation formed part of a larger plan, conceived and funded by the Global Ecovillage Network, to try and incorporate the environmentalists of the kibbutz movement into the Network. At that time I had heard about Permaculture, had read that the Global Ecovillage Network was in the process of being created, but had no direct links with either. My world in Israel consisted of attempting to persuade the otherwise straight kibbutz authorities, and other official bodies in Israel, to take on board an ecological agenda. After ten days at The Farm in 1995, I went back to Israel, co-founded the Green Kibbutz Group, and an informal Israeli Permaculture network. I went on to teach Permaculture, and write numerous articles and even a book about the subject. And I'm just one, short term, insignificant visitor to the place!

The influence of The Farm has been, and continues to be, immense. Literally thousands of people have passed through, some staying for years, others for a few days or weeks. Many of them have been profoundly affected by what they have experienced there, and have gone on to do significant things in other places.

Today The Farm is still there; it's still an amazing place, it feels good, and there are lots of good things happening in the community. It is the focus point and the stimulator of the ecovillage movement in all America. It is the home of the Ecovillage Training Centre where courses are held, and over the last decade, thousands of people have been trained in practical skills to do with gardening, building and community creation. Lots of other things have emerged from The Farm. The aid organization Plenty still sends funding, volunteers and know-how off to countries and communities in need. The Book Publishing House carries on bringing new books out every year. The Birthing Centre attracted national and international fame with its publication of *Spiritual Midwifery,* and the overwhelming success of its birthing rate after helping over one thousand births in its first years. The statistics proved that natural home-birthing was safer, more comfortable and more fun for everyone concerned. Another business on The

Taking the Permaculture Design Course

When I came to The Farm for the first time, in 1995, it was to attend a Permaculture Design Course. The whole trip had been arranged through the Global Ecovillage Network, and was part of a plan to introduce Permaculture and ecovillage ideas into the kibbutz movement and into the wider Israeli consciousness. I was the only non-American on the course, and most of us had not been to The Farm before. One thing that struck me was how safe the women on the course felt. They often talked about how they could not just walk around in the night where they lived, that it was dangerous, and that people were mugged and raped nearly every night.

That last observation was interesting to me, who came from the violent Middle East, but where women felt safe in most cities at night. Our violence was political, not social.

While we were at The Farm, there was another group there, Kids to the Country, a project designed to get kids from those violent inner city areas to spend a couple of weeks in the country, surrounded by natural things, and provided with good, healthy activities. Another example of how The Farm was integrating itself with the inner city areas that needed help to get themselves out of the sink position they had been drawn into.

Thousands of people pass through The Farm Ecovillage Training Centre, and most of them leave with favourable impressions of The Farm, helping it to integrate into a wider national and international scene.

Farm designed a geiger-counter for personal use which is now being sold to companies and government departments internationally, and the business is in the multi-million dollar bracket. The Farm has supplied Greenpeace with its electronic equipment, linking itself to what Stephen Gaskin calls 'our hippy navy!'

There is no doubt that The Farm has made an impact, nationally and internationally

Another way to measure success is to compare achievements with the initial aims. The Farm set out to create a better society. When I was there the first time, I was part of a group that had come together from all over America. I was the only foreigner. During the course, many of the women talked about how safe they felt at The Farm, something they felt the lack of in the neighbourhoods where they lived. For several of them, urban violence and personal insecurity was part and parcel of their everyday lives, part of their society. The Farm had gone one better: the place was safe for women to go for a walk by themselves in the evening without feeling afraid. Children could roam freely.

Whether society as a whole has improved because of the The Farm needs a more complex answer. Has society got better at all? Would it be even worse than it is without The Farm's existence? We are back to the subtle influences of the people passing through The Farm. Clearly the place isn't single-handedly going to save humanity, but with its variety of help programmes, reaching Central America and South Africa, its involvement with Greenpeace, its urban youth programmes and the already mentioned Global Ecovillage Movement role, it seems to me absolutely clear that the three hundred or so residents are making more than their fair share of positive improvements around the world.

My sense is that we can't apply the terms success or failure to communities in a 'yes or no, tick in the box' way. We need to look at the social ecology of the community, its relationship to other groups, to other sectors of society, to society as a whole. We need to look at subtle interpersonal relationships, we need to bring in to play a whole range of skills that are not generally recognized as scientific. Intuition I already mentioned, but I would add the skill of being able to perceive patterns, to trace the spread of ideas, to see if there are significant meetings between individuals, to look for those coincidences that strike us as more than just chance. Rather than isolate individual communities in sterile social laboratories, I want to begin looking at intentional communities as bearers of ideas, ideas that have as their intention to create positive change in the wider society. We are running a relay, and hopefully any sense of success will entail the handing on of the baton of social change, rather than trying to run the whole race single-handed.

The management of the community's finances is one of the contact points with the outside world. Influences can flow in both directions, with the ideals of the community making itself felt to others, but at the same time there is a danger that the financial practices of the wider society may penetrate into the community and influence inter-member transactions. Results can run both ways.

The Norwegian Camphill movement had a pension scheme for its co-workers, buying into a conventional one set up for local government employees. As a kind of social work institution, this seemed at the time to be the most appropriate. A few years ago it was found that not only was this costing us quite a lot of money, but that the actual pensions that people would be getting were not very high. In other words, it wasn't a very good deal. By this time the movement had established a very close co-operation with Cultura Bank, which was also inspired by the anthroposophical ideals of associative economics, and can be seen in a context of the general rise in ethical banking and investment impulse. A working group was set up, and after a couple of years of hard legal work, a suggestion for an internal pension fund was published, and got enough takers to launch. Today, regular payments are made for each co-worker who is part of the scheme, the fund is legally registered and managed by two committees. One decides upon the investments that the fund can be used for. No more than 20% of the fund can be allocated to the Camphill movement, but this is already a considerable sum. The 80% that is left is available for projects that are in line with the ideals both of the Camphill movement and Cultura Bank. The second committee decides upon the actual allocation of the pensions themselves, making sure that these are carried out in line with the quite complex calculations that the pension fund has set itself.

This is a good example of how a community can meet the complex requirements of the outside world, and create for itself such conditions that satisfy its needs and enable potentially creative change within society. But it can go the other way, though, and outside influences can impact negatively on the community.

When we joined Kibbutz Gezer in 1984, the advice that was being given to kibbutzim was to borrow money. It was said that conditions were right, and that it was economically advantageous to do so. The kibbutz movement was big, fairly strong, and had many decades of experience in co-operating with one of the largest banks in Israel. Who were we to argue with the experts? At that time, Gezer could point to statistics showing that it was one of the fastest growing kibbutzim in the country. It was clearly a mature community. The dining room was too small, plans were drawn up for a much bigger one, and work began in 1986. Membership topped the one hundred mark, but then began to drop. Founder members began leaving, many of them. At the same time inflation hit Israel in a big way. For a time no-one used the Israeli shekel; the US dollar was the only safe currency. The loans were linked to the dollar, but we could only pay back in shekels. Our loans spiralled away into the sky, reaching such dizzying heights that we knew we would never be able to repay them. It wasn't just Gezer, the whole kibbutz movement was in hock to millions of dollars. This story can be told elsewhere in more detail. I just wanted to give it as an example of an interaction between an alternative community and the wider society in the field of economics that had serious negative results.

One of the aims of most alternative intentional communities is social change. We perceive that society has problems, and dream up remedies. We are often not very impressed by the solutions offered by the establishment. The problems are mass problems, the solutions offered by economists and governments are usually mass solutions, and our ways of measuring success or failure are mass measurements. These solutions, tinkering with the money supply, adjusting the bank lending rate, encouraging investment or consumer spending, were methods worked out during a period when capitalism was enjoying rapid growth. We have now come to the end of that period, growth is no longer desirable or acceptable, and it certainly won't solve our problems.

We need to think out of the box. We can find an interesting pattern at the beginning of the industrial revolution itself. This was not a planned and co-ordinated restructuring of society, but rather a series of small, local initiatives, each one reinforcing the next and creating an overall pattern. Many were unsuccessful, but learning was born out of experience. We in the ecovillage and intentional community movement would do well to see ourselves as a similar series of enterprises. Friedrich Glasl makes just this observation in *The Enterprise of the Future:*

> We have to recognize that our own organization can only be successful if it sees itself as an element in a biotope, a whole inter-dependent network of relationships between different organizations and stakeholders.

In her much more detailed life cycle of organizations Margarete van den Brink sees in phases 5 and 6 the rise of moral principles, and greater awareness of the world as a place where our ideals need to manifest. Michael and Jane Luxford in *A Sense for Community* perceive this interface with the outside world in terms of quality control, but the end result is the same:

> Today, every organization, institution and community which is involved with health, social care or education has to face pressure over quality control and safeguards. We know that this pressure has to be balanced by something else, but what? Perhaps it is nothing more than the support we give to one another which will make the difference.

As communities mature, their interaction with the outside world becomes more complex and more intense. If creating social change is one of their aims, they need the internal strength of a sense of identity, a real culture, and they need the strength in numbers that will only come with support for each other.

12. The Poor Are Always with Us

There is a big difference between communities that practise income sharing and those that expect each individual or family to be responsible for their own economy. One system is not better than the other, though it might be argued that income sharing is more radical within the context of the capitalist and individualist west. Even though each system has its advantages and drawbacks, income sharing does not seem to appeal to as many people. I often get the reaction: 'Sounds great, but it's not for me!' when I explain income sharing to other people. I have been living in income-sharing communities for over twenty years now, and really appreciate how it frees one from some worries, but of course it replaces them with others. Security and self-development are the two main reasons why I prefer it. Sharing your income with others means that you can budget a good sum as an emergency buffer should there arise an unforeseen expense. Given that you trust one another, you can always ask for help if you need it. This could be medical, new glasses or dental treatment, or it could be the need to travel to visit a sick relative. Self-development is much harder to define or explain, and it sometimes sounds like self-flagellation. There is a serious challenge in sharing your life and economy with someone who seems to spend money like water, and in addition, you don't really like them. Working your way out of that tangle is really an opportunity for self-knowledge and self-development. Of course, if the only results are conflict and argument, it could be counterproductive.

Whichever system your community adopts, differences in status and access to resources will inevitably develop over time. Some individuals, by the very nature of their personality, will have more power while others have less. As the community matures, rotation of jobs will diminish, specialization takes over, and those who have a tendency to accumulate power will do so. It may sound depressing, but it actually isn't so bad. If a member is competent, trusted and open about what they do, then it could be very advantageous for the community if that person has the power to carry out good work. Should he or she get more money for their work? I'll leave that question for others to answer, but we all know that there are many ways of rewarding people.

The challenge is how to make sure that those less skilful in gaining access to wealth and status still retain the feeling of being in control of their lives within the collective. Poverty will never entirely be eradicated, even within a strict income-sharing egalitarian collective, but it can be severely limited, and made innocuous. On the other hand, we don't want to create people who are cloned automatons; we want, I hope, that each person develops his or her own self. We want individuality and self-realization. As people become individuals, they become different, some more, some less, powerful.

On the kibbutz, when we joined, we did not have the freedom to add on to our houses freely. If you felt you needed extra space, a room for your kids, for example, the issue had to go through members and building committees before anything was done. Larger families were given extra space, but it was a long procedure. Gradually there developed a feeling that this was all a bit old-fashioned, and that people should be able to do things themselves. I walled in our back porch, which was fine as long as there were big gaps open to the world. During the winter, however, the rain might blow in, or it got too cold to sit outside comfortably. I put in some simple doors and windows, and got a lot of criticism. My defence was that all the work had been done in my spare time, using largely found materials, and any purchases I made came out of my pocket money. The counter-argument ran as follows: I was using my skill to enhance my standard of living above that of other members. Another member, lacking the handyman skill that I had, would be forced to hire a builder to do the same as I had done, and that, we all agreed, was stepping over the line of acceptability.

The times were changing on the kibbutz while this argument ran, and soon afterwards a more open policy was accepted. It was interesting to see how standards of living began to diverge when members could bring in outside money and outside contractors to begin changing their houses. Today, with a total change from income sharing to a private economy, some people on Gezer live in unrecognizable and large villas, while others still live in the old style basic housing. But to be fair, even today, the rich are hardly rich, and the poor are not so badly off.

Cynthia Holzapfel, in an article about the change from communal economics to privatization on The Farm in Tennessee makes the comment that: 'There were, and always will be, individuals who have a harder time than others making ends meet.' This quote comes from *Voices From The Farm,* edited by Rupert Fike, a fascinating collection of articles written by members of The Farm, many of them describing their reactions to the changes that occurred there.

Living in close community is a continuous balancing act between freedom for the individual and the equality of the collective. One day a friend from a neighbouring kibbutz arrived to visit us on a motorbike. 'Don't tell anyone I came to see you on my motorbike!' He explained that his kibbutz had a car pool, like all kibbutzim, and a no private car policy. He had argued that he wanted a motorbike for the sport of riding it, just like someone else might want to go rock climbing or wind surfing. He was allowed to save and use his pocket money to buy one, but he was only to use it for sport and fun, not to go anywhere!

There is a saying within the kibbutz movement that the slippery road to privatization began with the first private kettle. In the early days there was just one kettle, in the communal dining room. If you wanted a cup of tea or coffee, you went there, and hung out with other members. It was all very jolly and communal. As prosperity increased members were allowed to have a small kettle in their private room, and could go back there for their coffee. The old die-hard ideologues argued that from then on it was an inevitable decline into the eventual closing of the dining room, and privatization of incomes.

Of course it doesn't have to be like that, but there does seem to be a pattern that, as the years go by, a strictly collective community will gradually loosen up, and things that were unacceptable in the early days inevitably creep in as OK. We'll see how far this goes when we look at how communities handle their old age. The big question for me is not the outward change, but how much the community can retain its culture and identity despite these outward changes. Ed Schein comments on this in the context of corporate culture:

> Organizations are started by individuals or small teams who initially impose their own beliefs, values and assumptions on the people whom they hire.

> If the founders' values and assumptions are out of line with what the environment of the organization allows or affords, the organization fails and never develops a culture in the first place.

These underlying beliefs, values and assumptions need to be there, underpinning our communal practices. Whether they are imposed by founders on those they hire, or are agreed to by the initial group that establishes a community, is not that important. If these core values cannot find some manifestation in the community, it will eventually break up. But change must be accommodated all the time, even within strictly egalitarian communities there has to be change. Charles Handy, looking at business organizations, sees this clearly:

> Much of the trouble in organizations comes from the attempt to go on doing things as they used to be done, from a reluctance to change the culture when it needs to be changed.

13. I Want My Own Room

The very first time we visited kibbutz, in 1972, child rearing was entirely communal. Children slept in children's houses and not at home. It was one of the hallmarks of kibbutz, and the subject of many studies, the most famous probably by Bruno Bettelheim, published as *The Children Of The Dream*. We came back for a longer stay in 1977, and then the subject of kids' houses was already up for debate. Some kibbutzim were going over to having children sleep at home with their parents, others were talking about it. When we finally moved in as member candidates in 1984 it seemed that most kibbutzim had gone over to having children sleep at home, and the kids' houses operated as daytime kindergartens. The most radical of the kibbutz movements, Kibbutz Artzi with about eighty kibbutzim as members, voted for kids to sleep at home in April 1987.

This was a major change in the kibbutz way of life, a watershed in its ideological history. The theoretical and ideological arguments for collective child-raising in purpose-built children's houses I find very attractive, but it has to remain just a nice idea. It wasn't such a nice practice. I know many second and third generation kibbutz natives who grew up in children's houses, and I can't remember a single one who would not rather have grown up at home with their parents. It may well have been appropriate in other cultures and at other times, and it certainly served the kibbutz movement well for a couple of generations, but the human consciousness is changing; we are becoming more individualistic, and the practice of such rigid collective child rearing is no longer viable.

This internally generated change had an unforeseen but profound effect upon the history of the kibbutz movement. Making the change to having children sleeping at home with their parents entailed constructing extra rooms for them. Up till then, couples had just enough space for the two adults. Some kibbutz families were very large indeed, but of course the children slept in the communal children's houses. So moving the kids home created a major building boom, costing lots of money and using lots of labour. Some kibbutzim did their own construction, others hired contractors. As explained earlier with the dining room on Gezer, this was the time that financial experts encouraged the kibbutzim to take up loans, followed shortly thereafter by the hyperinflation and subsequent dept spiral. There is no doubt that this children's room building boom added considerably to the financial problems of the kibbutz movement, especially as it was an investment which carried with it no returns. The increasing industrialization of the kibbutz movement, which was taking place at the same time, had at least the added capacity to generate additional income.

As a community matures, the need for private space increases. This can be the result of outside forces imposing themselves upon the community, or it can be internal changes. Both internally and externally, the pressure in a maturing community is to build more private space, to allow individuals more freedom.

In the early years of Camphill, houses were crowded and people often had to share rooms. It was quite normal for two or three villagers to sleep in one room. Sometimes there were villagers sharing with the children of the co-workers. When we arrived at Solborg in 2000, everyone had their own room, except for two villagers who shared, and they were both given separate rooms after a couple of years. It was during this time that we were engaged in a long dialogue with the local planning authorities, arguing that we needed more space in the village, and that this meant we needed planning permission for more building. The concern of the planning authorities was that we did not have capacity for more waste water treatment, their equation being that more space equalled more people, and that if we built we would exceed our planned size. Our argument was that as society imposed higher standards for the care of the mentally handicapped, we needed to expand, even with a static population.

According to Margarete van den Brink, in the early period (phase 2) everything revolves around the group as a whole, and individual and personal experience have very little significance. During the development of both the individual person and the group, by what she calls phase 6, 'the new community,' the individual is central and other members of the group put themselves

Children's houses on kibbutz

When the early kibbutz communities began to have children, there was a pragmatic decision to organize kindergartens and children's houses, so that the mothers could be freed for work as soon as possible after giving birth. This soon developed into a child-rearing ideology, with specially trained children's workers, whose aim was to produce 'the new human being.' At that time, during the 1920s, child psychology was in its infancy, and largely based on materialist worldviews.

Two generations later, it had become clear that this model of collective child rearing was lacking something, and during the 1970s and 1980s the whole kibbutz movement changed to family-based child rearing. In my own experience I cannot recall having met one single person who was reared collectively who would not rather have grown up living with their parents.

Whether this should be judged as a mistake, a failure, a brave experiment, or a product of its time is not up to me. I see collective child rearing as one characteristic of a youthful community, which inevitably will be replaced by more family-dominated structures as the community matures. On the way, however, there are many spin-offs, and these are interesting to look at.

I worked for many years in the day care houses of the kibbutz, working with children aged six to fourteen years old. These children went to a local school, came back around lunch time, we fed them, organized afternoon activities, and often extra things in the evenings. We operated as a kind of youth club. With the older kids there were house meetings, with a good deal of delegated responsibility and training for democratic decision-making. Around the building was an adventure playground, an ever-changing collection of old junk, which the kids built their fantasies with.

This was a really important social training, and the majority of the kids who went through this upbringing have turned out to be well balanced, skilful and sociable people.

Growing up together

Children who spend all day together from an early age, develop quite special relationships, not quite as strong as blood siblings, but a lot closer than normal. These relationships can follow them all through life, for instance I know of groups of kibbutz kids who have developed businesses together in other countries.

One of the things that I wish for is that my children become good friends with the children of my good friends. In community this comes about by having children more or less at the same time, so they grow up together in the same kindergarten and the same school class.

Going through child rearing together creates a really strong bond between people. This is reinforced when the resulting kids also grow up together. The kids of your friends are also friends amongst themselves. This has an added side effect when it comes to relationships between the adults. There is no guarantee that life in community is problem free. In fact, it's often full of problems, and many of them stem from conflicts between people. If your children are good friends it is likely that you don't want to destroy or impair their friendship and that you will keep your own conflict within the bounds of reason.

There are many reasons why children are a blessing for us in life generally, and in community there are even more good reasons.

at the disposal of the individual's inner development. It seems that this increasing privatization is an inevitable process within groups, even those who begin their existence with a very strong collective consciousness.

The Luxfords writing about Camphill, anchor this trend firmly in the spiritual epoch that we find ourselves in:

> It is a reality, that the ego forces are strong in our time and have to be so, in order that each person in freedom can follow his or her individual path. But we still need each other in order to develop and there I see a big gap, between what we want and what is happening. Because of the need to feed your own ego, people often become separated from each other, and are not able to reach out, to give back to the world.

This is also one of the insights of the 'Community' group which lies at the core of the Camphill network, though here I have only my own personal interpretation as an individual member of 'Community.' I see 'Community' as a group helping the individuals within it towards self-development, within the context of the fellowship. Later, when I look more closely at some spiritual aspects of community, I will develop this idea further. The idea that a group supports the individuals within it towards greater self-development is in itself a strong community binding factor. Freedom can be defined as the ability to choose to live in a system which one feels comfortable or satisfied with. Margarete van den Brink, writing about ideal-oriented organizations, has found that there is a lack of awareness in just this area:

> Here we come to the second problem area in ideal-based organizations. Within the institution or company there is usually little awareness or room for these personal and professional needs of the employees. Besides, the executives often lack vision and skills in this area.

Happy kibbutz family

Some communities have tried to replace the nuclear family with other structures. Kerista in California in the 1980s was a relatively short-lived attempt to create a group marriage. Kibbutz Degania in its first year, 1911 and 1912, would not allow marriage at all. The Shakers in nineteenth-century America practised celibacy. There have been other attempts also, and the one thing they seem to have in common is that as time went by, they were gradually replaced by the nuclear family asserting itself. Except for the Shakers, who just died out.

When we joined kibbutz in 1984, we were already a nuclear family, and had no intention of being anything else. The kibbutz had voted for children living at home a few years earlier, and we were happy with that. This picture shows us in front of our new kibbutz house in about 1986, balancing between living collectively and being a nuclear family. Every morning Ruth and I went off to work, the children went off to school or kindergarten. Breakfast and lunch were eaten together with our colleagues, most people came home around 4 or 5 in the afternoon, had free time together, and supper was eaten in the dining hall between 6 and 8. When the children were small, and needed help choosing and eating supper, they generally ate together with their parents. Then came a family changing event, around the age of five, maybe earlier, maybe later. The kids began eating together with their friends from the kids' houses. From then on, they were lost to their families; of course it was much more fun to eat with your friends, than to have to listen to all that boring grown-up talk.

It was not as rigid as all that. Many families ate together in the dining hall, some families ate supper at home. What the kibbutz gave was the opportunity to choose, to be either with your friends or your family, a safety valve that probably helped many families get through those periods when one member needs a little more space, a little privacy.

If we are looking for good combinations, I can find no better than this mixture of collective living, with the nuclear family as the basic unit.

In her analysis of seven periods of organizational development, Margarete van den Brink comments that a group may perceive itself as being at a different stage than it actually is. It seems that, as a result, ideal-oriented organizations often neglect personal and professional development of the individual employee or member, thinking they are already ahead of where they actually are. Who is right and who is wrong? Where is objectivity in this? When we come to assess human beings, it's fairly easy to assess their age, even if the birth certificate is missing. In most cases you can tell if a person is ten, twenty or thirty years old, roughly. When it comes to community, it's much more complicated. Solborg, where I live is nearly thirty years old, but only three co-workers have lived there longer than ten years, and none of them from the beginning. So how old is Solborg really?

Margarete also points out that a highly complicated and problematic situation will arise when some members of a group are in different phases from others, or the community as whole. When younger people join a mature community, for instance, problems might arise. This is usually easier in a larger community. Kibbutz has experienced this during its whole history, and has a clear membership process which was described earlier. Part of the function of this process is to bring the candidate 'up to date' within the community. There's also a numbers game. If two or three people join a community of a hundred, their impact is likely to be relatively small. But if two or three join a community of ten, their impact would necessarily be much greater.

To round off this look at community in its mature phase, I found a fascinating comment from Nancy Foy, looking at organizations:

> Changes seem to come in cycles. Berg suggests that periods of revolution are natural in an organization, preceded and followed by periods of stability. As in mid-life crisis, the potential for rapid change is probably lurking inside most organizations, simply waiting to be triggered after they reach a certain point. Sometimes the change is healthy, and sometimes it seems disastrous; sometimes it's planned, and sometimes it's something that wells up out of the organization of its own accord.

This confirmed something that I have also observed, that there seems to be some inner dynamic that governs the community's ability to change. The potential for change is indeed always there. But what is it that triggers? What forces are at work within a group of individuals that in some cases produces healthy growth and the ability to respond to change, and in other cases results in totally dysfunctional community decisions, conflict, and the break-up of the collective? As we go further, looking at old age and beyond, I hope to take up these puzzling questions.

Part 3

Old Age

The transition from the youthful pioneering phase to the mature phase is much easier to see than the transition from maturity to old age. In this relatively simple tripartite division, maturity dominates; it will last longer (hopefully) and settle into a kind of steady-state condition. Old age is the end of maturity. We'll look at real old age, senility and death, in the last part of the book.

Nancy Foy in *The Yin and Yang of Organizations,* raises the idea of these same three phases:

> Finally it would make sense to raise the question of whether companies, like the individuals within them, go through these phases (youth, maturity, old age), resolving questions of identity and destination.

There are some characteristics which we shall look at, beginning with increasing specialization and the privatization of the economy. Once a community has several generations of members it is without doubt both well established and into old age. Similarly, once a community movement is set up, the members can consider themselves to be well beyond youthfulness and far enough into maturity to be old.

There are a couple of images which I have had before me as I worked on this material. One is the image of the river in its landscape, again divided up into youthful, mature and old phases. The other is the image of the tree, the seedling, the mature tree, and the old stately tree. This is useful as a Permaculture exercise in matching patterns found in nature to systems that we design for ourselves. I have found some associations here and there, and leave it up to you, the reader, to come up with further examples.

You may also like to ponder on the following statistics while reading through the next few chapters. They are gleaned from Chris Coates' book *Utopia Britannica*. There he describes 350 groups that came into existence in Britain between 1600 and 1945. He then looks at the lifespan of these communities and finds that very few lasted more than fifty years:

Age/years	Religious	Secular
Less than 2	20.5%	9%
2 to 5	27.25%	30%
5 to 10	13.5%	18.5%
10 to 25	8.25%	15.5%
25 to 50	12.5%	17.5%
50 to 100	11%	7%
100+	7%	2.5%

It may look as if communities inevitably dissolve and fade away. Don't be dismayed; now read on ...

14. The Professionals Take Over

There is a danger that some of the founder members of communities often look back at those first golden years and lament the lost innocence when everyone did everything. Like we all do when we look back at our own youth, and reminisce about the days when we could run a mile, climb a tree, stay up all night, and had lots of energy and tremendous strength.

There is a price to growth and development. We may move on to become more mature, but we will have to leave behind the things of youth. The price of not growing, however, is even higher. A plant that never develops beyond the seedling stage will wilt and die. People, and communities, need to move on, to grow and develop. It's OK to mature, don't knock it.

One of the symptoms of this in community is greater specialization. Saadia Gelb spent a lifetime on kibbutz, and wrote a book of reminiscences and observations. Much of what he writes about is about change:

> In due course intellectuals were attracted to kibbutz life. Many joined and were integrated into the community. When, however, a lawyer wanted to join, it was not quite clear what to do with him/her, but they eventually became teachers and farmers and what not.
>
> No one today asks what to do with lawyers. They are sorely needed to:
> ... read fine print clauses in bank loans
> ... scrutinize industrial contracts
> ... prepare export agreements
> ... defend the kibbutz against unhappy ex-members
> ... and a multitude of other activities.
>
> So if you know any good lawyers who want to live the simple life of the kibbutz ...

As I pointed out in the very first chapter, one of the characteristics of the pioneering phase is that everyone does every job, or nearly so, and that people with specialized professions very often find themselves doing unexpectedly different things. As the community grows and matures, the increasing complexity of running a major enterprise will require a greater degree of specialization. Gone are the days when everyone could turn their hands to everything. This should be accepted as part of the maturing process that a group goes through. By the time the community settles down to old age, specialization and professionalization should be well-established features of the group. Young people growing up within the community might spend some time during their school years and maybe a little after, moving round doing different jobs, but they will probably be aiming at some kind of specialization, and in most modern communities, will most likely study or train into a recognized and needed profession. If they stay within the community, they will be useful, contributing members. Should they decide to leave, they will have a secure profession to follow in the outside world.

In a modern society, certain skills need to be authorized in order to comply with laws and regulations. Electricians, architects, and engineers will need to have certification. If members of the community don't already have these qualifications, they can be hired in. Another option is that the community may send members off to learn. Inevitably a group of specialists will be created, and issues of status, rights and special treatments will arise. In mainstream society these are often high status and highly paid professionals. Architects or lawyers who are members of egalitarian communities such as kibbutz or Camphill will inevitably have a good deal of contact with their professional peers, and may find it hard to accept that within community they are poorer. Of course they have lots of other, less easily measurable advantages: comradeship, security and access to resources. But these don't look as good as a new Porsche!

Another danger is that as the more competent members of a community increasingly take on more responsible tasks, their status often rises, while those less skilful end up with the lower status tasks. Specialization can become an issue that easily leads into conflicts. In the classic kibbutz, and in Camphill, this was not so much about money, as the material rewards, pay, housing, and so on, were not related to the work that you did, and the toilet cleaner got the same as the financial manager.

Planning the infrastructures

In the good old days, it seemed that you could just move out to the country and start building community. Today things have changed, and there are bureaucrats to satisfy.

I recently heard about a community that is trying to negotiate itself through the planning application. They have to satisfy not only the Planning Authority but also Building Control, Environmental Health, Fire Regulations, Residential Care Home Standards, Social Services and the Department of Transport and Highways. Each of these is laying down how things must be done, while one of the main reasons for trying to set up the community in the first place is to demonstrate how things could be done in an alternative way.

Within the building world, insurance and fire regulations demand that your building work is carried out or checked by registered plumbers, electricians, builders, architects and engineers. For those of us wanting to use natural, scrounged or recycled materials, building regulations often permit only registered materials. Of course, it is important that eco-community infrastructure is carried out properly. Experts are needed, and the link up to public services such as water, sewage and electricity, if it is desirable, has to be done in a proper way. We might demand a similar attention to details from heavy industry and big business, specially the ones who pollute seriously.

My very first impression of kibbutz was sitting in the dining room of kibbutz Ginossar in 1972. We had just arrived and were taken there for a cool drink and to sort out what we were going to do.

'See that guy over there, cleaning tables?' said our host. 'He's Yigal Alon, member of the Knesset and a cabinet minister.' It turned out that Yigal Alon was a member of the kibbutz, and that while the Israeli Parliament, the Knesset, was in summer recess, he was at home and worked in the dining room. I was very impressed! It's not often you find cabinet members taking summer jobs cleaning tables in cafeterias.

So rewards are not always in wages and money. There are things such as power, status and side effects. Those who specialize into management often wield considerable power. If you can talk well, and present the community favourably, you may find yourself quoted in the media, appearing on TV or on the radio, or being asked to write articles or books. If your job is to represent your community in meetings with other bodies, planning authorities, funding organizations and so on, you will inevitably find yourself travelling to meetings, maybe going to conferences, sometimes to other countries. All these activities are potential sources of resentment amongst members who don't wield power, don't get their name in the newspapers, or don't get to travel as part of their job.

Then there are issues of career and ambitions. Competent individuals tend to take on more responsibility. But how much can you get in a small community of forty or fifty people? If your village is part of a movement, a federation or a network, it seems obvious that for the competent managerial type, the next step will be for them to move into some job co-ordinating the different communities that make up the network. That will take them even further out of their home community. The kibbutz movement had central offices in Tel Aviv, and members who worked there would commute into town. From the more remote places like the southern Arava, down near Eilat, there were regular commuting flights morning and evening. A year or two ago the Camphill movement here in Norway set up an office in Oslo to co-ordinate activities, and one member now commutes regularly into town. Sometimes some of these highly professional manager members leave, either carrying on in the same jobs as consultants, or moving into other industries or organizations, taking their managerial skills and experience with them. The community is faced with a new problem. How does it retain its experienced and valuable membership? This is not a problem confined to communities. Nancy Foy, writing in *The Yin and Yang of Organizations,* found that mainstream corporations are facing the same problem:

> Ultimately, the company's vitality depends on its ability to get and retain (through mid-life years and beyond) good managers, so the joint investment in an individual's development makes sense. An organization that contains a healthy population of these serene seniors, whose well-integrated lives can create potential patterns for others, has a better chance of maintaining balance, not only in ages but also in attitudes.

On Gezer we had an unfortunate series of experiences in the kibbutz car and tractor workshop. During the sixteen years that we lived there, the kibbutz paid three times for members to train as tractor mechanics. One after the other. This cost a lot of money, and the three members spent quite a bit of time going to school, eventually getting their certification. They all left the kibbutz either while they were studying or after they qualified. One after another. One of them did offer to carry on working in our garage, at full mechanic's pay. Many members of the kibbutz were very unhappy about that offer; after all, we had just paid for him to qualify. After these kind of experiences, some communities have begun the practice of asking members who want to train to sign contracts, that they will commit to staying on for a certain length of time after they qualify. This only partly solves the problem. It can be pretty uncomfortable to have someone stay in community because they are forced to. Ideally we would like people to stay in community because they want to.

One area where continuity is really important is in the maintenance of buildings, and the grounds of the community. With a large village this can be quite a daunting task, keeping an awareness of all the services that a modern community requires: water, electricity, sewage, telephone, and so on. Underground cables and pipes can be very troublesome when new building or other work needs to be carried out. In many communities there is not a complete and updated documentation of digging work that has been carried out over the years, but usually there is someone who has been involved in building work who knows where all the pipes and cables have been buried. This can be a member of long standing, or an outside local entrepreneur who has been hired by the village to do regular maintenance work over the years.

It's useful to bear in mind that the most important aspects here are not these material considerations of who knows where the sewage pipes are, or who has been trained as a tractor mechanic. Underlying these are deeper questions of how to retain the values and visions that originally inspired the original founding of the project. The Luxfords saw this clearly in their travels round the world of Camphill:

> We are faced with a great challenge. The Camphill Impulse created so much, and the Community brought about the Movement. What used to be taken for granted hardly seems to exist any more. This can be a good thing as the holding of the Bible Evening, or to have a common study and the like, have to be consciously decided upon anew every time. The challenge is to become re-convinced about the value of certain principles and intentions, individually and communally. This has to be done in the twenty-first century, with a legacy of over sixty years of the history and traditions of Camphill. It is going to require a tremendous spiritual and cohesive effort to regain a sense for community in the age of organizational structuring and systems analysis.

A community is on a journey: it is a group moving through time, and in intentional community there is often a goal, usually to do with social renewal. Just as on a journey you will stop and refer to the map, check the compass, and make sure that the direction is still right, so might a community take time out to check if the course is being followed. With increasing specialization, whether forced on the community by outside factors such as quality control or regulations, or naturally occurring within the community over time, there can be a tendency for individuals to build their own castles, to keep their heads down and concentrate on their own limited fields. What might happen is that they begin to identify their particular specialization as being the core task of the community. Within Camphill I found a good illustration of this while studying the organizational structure.

I identified two main trends, overlapping each other. One trend or impulse consisted of working with villagers in Social Therapy. This was often voiced as 'the villagers are the centre of our lives,' or simply as 'our work.' The other trend consisted of creating an alternative society. This would come to the surface when co-workers talked about 'the fellowship,' or 'realizing a threefold society' which we shall look at more closely in the next chapter. Sometimes they conflicted, or one tended to overshadow the other. This conflict may have been more apparent then real. In my own personal opinion the two impulses are parallel rather than in opposition. In fact, Camphill gives us the possibility of doing both at the same time, and this is one of the features that gives the Camphill movement such strength and vigour. There are plenty of alternative communities around where people can realize new forms of fellowship without having to come into contact with the mentally handicapped. And there are many institutions based on anthroposophical social therapy where co-workers can go home to their nuclear families at the end of an eight hour work day and live otherwise perfectly 'normal' lives.

Only in the interests of analysis, and as a tool for understanding, does this dual impulse idea have any value. It is only in Camphill (though there are other Camphill inspired communities which are not officially part of the network) that these two impulses manifest together in this way. When in perfect balance, this gives the Camphill tradition a robustness that has carried it through over sixty years and into over twenty countries throughout the world. It combines 'doing good work' with 'building a bright future.' Even though older established Camphill communities may be suffering from lack of co-workers and wondering what their 'vision' might be, new Camphill villages are founded nearly every year, and young people flock to experience this amazing phenomenon. We at Solborg turn away many young people every year for lack of space!

When these two are not in balance, however, conflict can rear its ugly head. We can imagine this in two extreme possibilities: co-workers with decades of experience and who relate wonderfully with villagers bypass the democratic process, or have no regard for financial accountability. Other co-workers spend all their energy on administration, travel endlessly to seminars and conferences, and seldom come into more than fleeting face to face contact with villagers. If this happens, even in a much milder form, it can easily be the trigger for conflict within a community.

Recognizing that Camphill straddles these two impulses may help some of us resolve some of our conflicts. In fact, if we can see conflicts as indicators that the two impulses have come out of balance, we may be grateful to the 'conflict' and use it as an opportunity to get things right again. If we change some of the words we use, and talk instead about 'in-balance' and 'out-of-balance' this might be enough in itself to defuse many situations and point us on the right road.

I tried to portray this graphically, in order for it to be clearer. If we can envisage these two trends as overlapping circles, we see that within each circle there can be many types of communities and projects. Only where the two overlap can we find the unique phenomenon of Camphill. This is why Camphill co-workers sometimes talk about those with mental handicaps as being our guides into a new type of society.

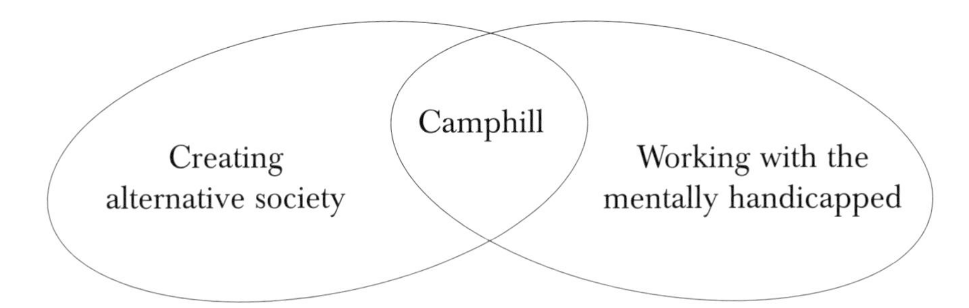

The most common strategy for a mature community to cope with increasing specialization is to institutionalize it, wrapping it in a framework of regulations and rules, often based on established habits.

Ed Schein observed a similar phenomenon in his study of corporate organizations:

> With growth in organizational size, people can no longer remain functionally familiar with others, so they have to resort to more formal processes of contracting, monitoring each other, and in general substituting processes and procedures for personal contact.
>
> As deals have to be negotiated with strangers, trust levels erode, and political processes begin to replace teamwork in pursuit of common goals. The sub-units become power centres, and their leaders become barons with an increasingly local agenda.
>
> For as long as the founders or founding families retain ownership and control, they can function as the integrating force and use some of the basic assumptions of the culture as the primary integrating and control mechanism. Charismatic founder owners can continue to be the glue by articulating the values and principles they expect organization management to follow.

If we can substitute the founding family with a regular reappraisal of the original visions and values, as I mentioned earlier, then we may be able to keep a perspective, even within the gradually specializing and in a sense fossilizing community.

Hopefully keeping some youthfulness, even in old age.

15. I Want My Own Money

As communities and groups mature and grow older, there seems to be an inevitable tendency towards individualization. This takes many forms, but few are as sensitive as the sharing of money. A successful collective can be torn apart from having too much money if the individuals then become infatuated with the prospect of private gain. On the other hand, a continuing feeling of collective poverty can be just as destructive. That was clearly one of the contributing factors to Gezer's drastic move to individualization. As an economic entity, the kibbutz was grossly mismanaged, opportunities for profitability were missed, time and time again, and the collective debt built up during the 1980s, by building the dining hall and the extra housing for children, created a situation where the kibbutz was poor. At the same time, individual members could see that by going out to work in paid jobs, they would be much better off financially, which is exactly what happened. They can hardly be blamed for taking that road.

There does not seem to be a correlation between how wealthy or how poor a community is, and to what extent it adheres to collective principles. Both the lack and the abundance of wealth carry with them dangers of conflict. But in both cases there are also opportunities for realizing community ideals.

Camphill in Norway is very strong financially, having negotiated a pretty secure arrangement with the government department responsible for the handicapped. Some members may bemoan the 'slide into individualization' that seems to them to be taking place here, but within the broader context of community the so-called slide is pretty small. Twenty years ago every member of the economic fellowship had to account for his or her spending to the group. Today each person asks for a regular monthly sum to be paid automatically to cover out-of-pocket expenditure. The rest goes into a buffer fund, and extra expenses need to be specifically asked for. I find it hard to see this as a revisionist slide into capitalism!

Individual personal wealth can certainly tear apart the fabric of community. In the late 1940s and early 1950s, many Holocaust survivors joined kibbutz in Israel. As things settled down after the

Second World War, the Holocaust was seen to be the monstrous crime that it was, and the survivors were given compensation in the form of payments from the German government. For most of the survivors, who did not live on kibbutz, this provided a much needed financial security, and at least some material compensation. For kibbutz members, who already had the security of the collective, this raised a problem. Shouldn't the money go to the community? After all, kibbutz is income-sharing. But these were in many cases highly traumatized people, who had a serious need to keep the money themselves. In many kibbutzim this led to serious conflicts, a battle between strongly held ideologies, and individuals with a deep need for sensitive treatment.

Perhaps it is just in this area of money, that the process of maturing to old age encounters some of its most serious problems. One attempt to solve this in a wider context was suggested by Rudolf Steiner in his book *The Inner Aspect of the Social Question:*

> How must human beings themselves be interrelated, so that they will be free members of the social organism and be able to work together for what is right and just?
>
> This way of thinking makes its appeal, not to theories or social dogmas, but to human beings. It says: Let people find themselves in the environment of a threefold social order, and they will themselves say how it should be organized.

The social aspects of anthroposophy are most developed within the Camphill Villages, where the threefold division of society is regarded as a basic tool for modeling the life and structure of the community. This threefolding was presented by Steiner in lectures during the last part of the First World War and the years that followed. He based his thoughts on his study of the development of European society over the preceding centuries. In England, he saw the industrial revolution as the modernization of economic life, leading to demands for fraternity, the development of trade unionism and labor party politics. In France under the French Revolution he saw a change in the legal life leading to demands for equality, and in Middle Europe (later unified to become Germany) changes in the spiritual life leading to demands for liberty. Steiner traced how these three great ideals, of Fraternity, Equality and Liberty had been corrupted by the rise of nationalism and the development of the centralized nation state. After Steiner's death, König further traced how this led to the insanity of Nazism, fascism and state communism.

Threefolding was presented by Steiner as a way of rebuilding Europe after the disaster of the First World War, but his ideas did not gain credence, and the ideas were largely dormant until taken up by König in building up the Camphill communities in the 1940s and 50s. It works like this:

— We worship and philosophize, educate, create music and art in the *spiritual sphere.* Here we need our freedom to develop ourselves.
— We decide amongst ourselves in the *sphere of laws* and rights, and need to regard ourselves as equals, with equal rights.
— We work, produce, buy and sell in the *economic sphere,* and need the fellowship (brotherhood and sisterhood) of looking after each other, not necessarily as equals, for clearly, some have more capacity and some have greater needs.

These three spheres are always with us, they are not determinants of how we should or might behave, but an attempt to make sense of our every day lives and how we come together as human beings.

So thinking from this perspective, what is money? A medium of exchange? A measure of work? A value? Personally I find it helpful to think of money in these three ways: as everyday living expenses, creative investment, and spiritual gifts. We try and understand it as these three separate things, though we use the same currency for each one. When I go shopping for food, or pay the electricity bill, I am covering the running costs of everyday life. This is one kind of money, living expenses; it pays for our everyday needs; it comes and goes pretty quickly, with a high turnover. It's like the rain that falls on the hills and flows down the little streams which are swollen during showers, but dry out again during a dry spell. Sometimes I have lots of money in my pocket, other days it's pretty empty in there.

If we can save some of the money from this everyday turnover, we can put it aside and invest it in new, creative projects. We can start new enterprises, build houses and expand businesses. This is investment money, and perhaps we might envisage this as a lake that our little intermittent mountain stream flows into. This lake will act as a buffer and a storage. The stream flowing out of it will have a much more regular flow, and the lake itself will support a more stable and richer ecology round its shores. In the same way a good business will generate a steady income.

The third kind of money is gift money, something which raises the quality of money into the spiritual sphere. Gifts are given freely, creating connections that might otherwise not happen. They can be thought out carefully beforehand, or be responses to requests, or completely random choices. This money is like water which evaporates from the oceans, drifting across the world as clouds. Eventually it will fall as rain, irrigating crops and forests, freely distributing the gift of moisture that is necessary for life.

In community, we as individuals can share our economy, and each of these three spheres then becomes much larger and more robust. Our total running costs might be much higher, but by bulk buying and direct deals we can reduce unit costs. Our investment potential is greater, and of course our gifts can be more substantial. However, as our community matures and grows older, and the pressure of individualism increases, so may the feeling that each member wants greater personal decision-making power over these things. Interests then begin to diverge.

Most of the co-workers who have made a commitment to living and working long term in Camphill have formed an Economic Fellowship where we share our income. This means that we put all our earnings into one account, and meet regularly once a month to discuss how to parcel it out. The basis is one of equality: we start off with the same amount of pocket money, but from then on things become unequal. I may have three kids needing schooling, my neighbour may have five, and the couple who live a few minutes away, and are responsible for the third large family house in the village, maybe have no children. Clearly we three families receive quite different amounts of money in order to cover the cost of feeding, clothing, schooling and dealing with our kids.

The simple egalitarian way to deal with money issues is to invoke equality, to give each person the same, and tell them to be responsible for themselves. Another way, much harder, but much more educational, is to bring up our differences, look at them, and make sure that our very different needs are met. Living with equality is relatively simple, even mechanical, everyone gets the same, everything is fair. But as human beings we are all different, and have varying needs. Accepting that and living with it is much harder, but opens up the possibility of learning more about other people, and our own response to them. Loving people who are nice and friendly is no big deal, fine in its own way, but not likely to challenge you. Learning to love your enemies is much more challenging, but more likely to make you yourself grow as a human being. When we share our economy, the challenge of living, working and loving someone who is greedy, lazy or 'not nice' throws you back into yourself, forcing you to take stock of your own prejudices and expectations. But it's not easy; it brings you up to face all the greedy, lazy or 'not nice' features of your own personality!

As communities grow older, you might get tired of this constant learning situation. After twenty or thirty years of this, it's quite understandable when someone says that all that theory is fine, but just give me my X amount of euros or whatever each month, and let me get on with my life. It's also quite understandable if a second or third generation member just wants their fair share of the surplus, and argues for an equal division and distribution of the profit. This is not necessarily a cop-out; it's just a sign of a group that is growing older. In every case that I have come across of a radical pioneering group keeping their cash in a basket in a communal space, to be taken out by individuals according to need, this has stopped after a period.

Keeping your cash in a basket seems a minor example when compared to communities making the decision to change from a collective to a private economy. I have mentioned The Farm and their big changeover, and how the kibbutz we lived on went through a similar process. Another example of this happened in the Amana communities in the 1930s.

I'll have more to say about the Amana communities in Chapter 18 dealing with old age, but here I want to use their change from 'communism' to a private economy as an example (you can read

What's left of Amana?

The sign in this picture carries with it two completely different experiences.

Are we who live in community doomed to end up forgotten, or as a display case in a museum? The more I have studied the phenomenon, the more it becomes clear to me that communities don't last that long. Very few last more than a century. I can only think of the Hutterites who can claim a continuity longer than two hundred years. Here at Amana their collective economy and society finally broke up formally in the 1930s. Now all that is left is a museum, a few books, memories amongst the older residents, and a visitors' village which is more like a theme park, complete with little Amana dolls that you can take home as souvenirs.

For many people committed to alternative community living, this could be a doom-laden scenario.

But is it? I come to visit Amana, nearly eighty years after the collective broke up. I look around at the buildings, I hear the German accents of the older people, I look at the cultural artefacts collected in the museum, and I buy books about the history of the community. I am at a conference devoted to communal studies, convened in Amana in order for us to be able to study this phenomenon. The ideas still live on, handed down to another generation. Who knows where these ideas and inspirations will travel?

In the end, inspiration wins out over depression. I come away interested in Amana as an example of an attempt to build a better society. I see community in a new perspective. Back home I told people I had seen a glimpse of the future, that when our community folds our ideas will live on, ready to inspire others by our example.

I even wrote a book about this. You are reading it now.

a more detailed account of this changeover in Peter Hoehnle's account called *The Amana People).*

The Amana Colonists settled in Iowa, USA in the mid-1850s, after a history of communal living in Germany in the beginnings of the same century, and an interlude in New York State a decade earlier. They trace their origins back to 1714, amidst the Radical Pietist impulse and the Community of True Inspiration. By the end of the nineteenth century they had established seven villages in a valley covering 26,000 acres, which had about eighteen hundred inhabitants. They were very religious and completely communal, eating in large dining rooms, bringing up their children in kindergartens, with very little private money or possessions.

During the first decades of the twentieth century it was increasingly difficult to keep the community separated from the rest of the world, and the economy gradually deteriorated. In 1932 the Amana community changed from a completely collective economy to a privatized one. From eating in communal kitchens to each family cooking and eating for itself. From working freely for the communal good to earning 10 cents an hour. After the Change, members gradually began buying their own homes.

At the time of the Change, it was generally the younger residents who wanted to decollectivize, while many of the older ones viewed this new privatization with a good deal of trepidation. Today the Amana Colonies retain a distinct culture, the economy is thriving, tens of thousands of tourists flock to experience their cooking, their handicrafts and their beer. They may have lost their communism, but they have not lost their identity.

This change, from communism to a private economy, is not uncommon as communities progress from maturity into old age, indeed, I would call it a characteristic of old age. Mark Holloway, writing about American communities of the nineteenth century in *Heavens on Earth* makes the interesting observation that communism, when combined with religion, was pretty successful.

> The Ephratans, the Shakers, and the Inspirationists
> have lasted longer than one hundred years; the
> Rappite community lasted almost a hundred years;

> the Zoarites and Icarians, fifty or more; Bethel, Aurora, and Oneida, more than twenty-five years. With the exception of the Icarians, these were all religious groups, for whom communism was not an end in itself, but a means of perpetuating a religious way of life.
>
> Ideal communism, as attempted at New Harmony and among the Icarians, was a disastrous failure in the one case and less satisfactory in the other than the communism of the religious communities. Absolute, ideal communism was a failure. The expedient communism of the religious groups was successful.

I wonder if we can learn something from that?

16. The Next Generation

The one person who stayed at the centre of The Farm, from its early days of the bus convoy right through to today, is Stephen Gaskin. Book after book appeared, the legend grew, the stories abounded. His skill lay in his being able to talk to large audiences, to crystallize the dreams and visions of a movement, and to carry on attracting and inspiring people. While I was visiting The Farm in the summer of 2004, I went to see him one rainy afternoon.

I found his house after driving around for a bit. There were houses dotted about in the woods, not signposted very well, so I had to call in at one or two other places first. Comfortable houses, a big improvement from the buses and army tents of the pioneering days all those years ago. They were large houses, but not luxurious or palatial. They were clearly home-made, often from used or found materials. Stone, timber, lots of wind chimes, colourful decorations, rubber boots around the doors, sheds and garages full of useful bits of junk, bicycles, old cars in various stages of either decay or repair, stripped down tractors. An unpretentious, working community.

Stephen lived in a large rambling house, children and grandchildren ushering me through living rooms and kitchens, up and down steps, to his bedroom, where he was lying on his bed, talking to another visitor who had got there before me.

'Best place to hang out with people is on the bed,' he greeted me. 'A habit we picked up in the buses, where there was so little space!'

Stephen is not a guru, just a really nice guy who seems to be able to get on with anyone, finding the right connection, able to talk the other person's language without descending to idle chitchat. So we talked about a wide selection of subjects, not least the up-coming presidential elections. Stephen had run for president in the past, and built up a relationship with Al Gore and Ralph Nader. We talked about science fiction, one of Stephen's favourite subjects. And of course we talked about community. The conversation was about as rambling as the house he lived in.

The rain cleared and Stephen invited us over to see Rocinante, a project he has been working on for the last decade or so, a retirement community within The Farm, a hospice for ageing hippies and others. Rocinante is not properly part of The Farm, but a smaller adjacent piece of forest, bought separately. The idea has developed along the way. When I first heard of it many years ago, it was envisaged as being a combination of the famous Birthing Centre, a place where babies could be born, combined with a place where elderly people could face the last days of their lives in a safe and supportive environment.

One of the big achievements of The Farm was the establishment of the Birthing Centre, which developed out of Ina May Gaskin's profession as a midwife. She was Stephen's long term partner, and another pioneer of the bus trip days, where she was already handling births. By the time The Book Publishing Company published *Spiritual Midwifery* in 1977 the Birthing Centre had handled over a thousand births, enough to show statistically how successful they were. The idea at the core of Rocinante was to build a properly designed Birthing Centre, combining it with the hospice and care centre for old people.

The reality has not turned out quite like that. The Rocinante that I saw was a collection of hand built cabins spread out in the forest, not all of them finished yet. A couple of people already lived there, others were about to move in, but as yet there was no centre to really look after those that needed help, neither had the birthing centre managed to get together the funds to build a special place for itself there. But Stephen was optimistic. The seed had been sown, now it was the time to be patient, to see whether it would sprout, whether enough people were willing to make it a reality.

It used to be the custom on traditional Norwegian farms that the grandparents looked after the children. Often a detailed contract was drawn up when the time came for the oldest son to take over the farm. This gave the old people the right to live in a smaller house, and an agreed upon amount of produce from the farm: firewood, food, clothing, etc. While the farmer and his wife worked the farm, the older folks often looked after the children, keeping them out of harm's way, and teaching them the skills they needed for when it became their turn to work as adults. A kind of leap-frogging of the generations, allowing those in their prime to be able to devote themselves fully to their work, and at the same time ensuring that the traditions were passed on and kept alive. Bringing together the oldest with the youngest.

Birth of a dying centre

One of the activities at The Farm, right from the beginning, was an active midwife service. This was needed by the community, which grew quickly, and they had a lot of births. It was also offered to outside people, especially to young mothers who were contemplating abortion. This was essentially in line with their pacifist, non-violent, vegetarian practices, based upon showing respect for life in their daily actions.

As The Farm grew older, so did the people living there, and the idea of a place to retire for older people arose. Stephen Gaskin spoke about the idea on his numerous tours. He had enough resources to create a space for such a retirement place, and waited for others to be sufficiently inspired to make it happen.

When I visited in 2004, the dream had not yet been realized, though some cabins had been built in the woods at Rocinante, and there were a couple of people already living there. Money to build a properly designed birthing centre, combined with a care facility for the elderly and dying, had not yet materialized. There was a faith in the vision.

The idea of combining these two activities has tremendous power. There is something mysterious and real about our coming into this world, and our departure from it. Those of us who witness birth and death will recognize that in these events are doors that give access to how we see our existence, to what is the meaning of life.

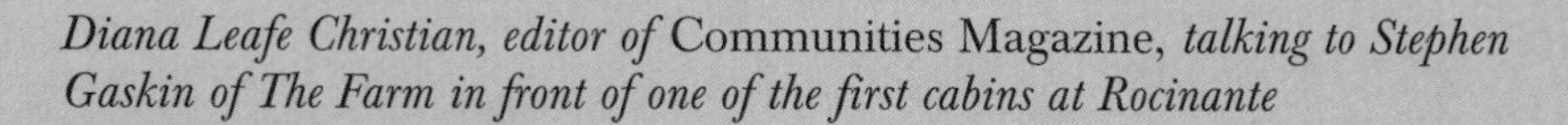

Diana Leafe Christian, editor of Communities Magazine, *talking to Stephen Gaskin of The Farm in front of one of the first cabins at Rocinante*

Rocinante was also meant to close the circle, bringing together the beginning of life with the end of life. There is a deeply mystical aspect to this. We can ponder upon the questions of where we come from, and to where do we go. Coming from somewhere, what do we bring with us, and in our going out again, what do we take from this life? Anyone who has been present at a birth will recognize how deeply significant an experience that was, as will anyone who has been at a death bed.

No community is really complete until it has been through both, until there have been births and deaths. A few weeks ago I received a newsletter from Camphill Svetlana in Russia. They had had their first death a few months previously: 'As the first death to affect us directly, it also leaves the Village changed in another way. Suddenly, everything is different. The cycle is complete, and we are at last a true Village.'

So we can see that another characteristic of the truly mature community is the presence of different generations, and the experience of both birth and death. Looking back at those good old pioneering days, we will find in most cases that the group was composed of younger people, with a relatively small spread of ages. If the community remains small, and is unable to recruit regularly as it ages, this will result in a kind of generation bulge rising up through the years. This could lead to serious problems later on, when most of the members are older, unless there are well worked out pension schemes or other safeguards against the time when the members can no longer work full time.

Some communities do everything they can to encourage their children to carry on living and working within the community. Others do not perceive this as of any importance. This particular aspect is one area of fundamental difference between Camphill and Kibbutz, and indeed is perceived differently in many other communities.

In the long distant past it was probably natural that children stayed near their parents, often inheriting land, buildings or professions, and carrying on the family line in the same location. Many communities regard this as natural, and are built on the premise that generation will succeed generation. Amana is a good example, as is kibbutz and other well established community movements such as the Hutterites and the Bruderhof.

How much pressure is put on the children to remain varies. The most intelligent way of dealing with this seems to me to be the way Bruderhof encourage their children to go to standard public secondary schools, and spend some time out of the community before making a conscious decision to come back in.

Other communities regard themselves as radical alternatives to straight society, demanding of their members a definite commitment. Children growing up there may not have the same ideals, and there is no guarantee that they want to carry on the same radical direction. The most extreme examples of this must be the Christian monasteries where celibacy ensures there are no children, and adult recruitment is the only way to keep the community going. In this they have been amazingly successful, with some monasteries tracking over a millennia and a half of continuous life. Monasteries really deserve a separate study; they seem to behave in fundamentally different ways from the communities I have been concentrating on.

Camphill can be seen as a holy order by some. There are indications that Karl König had ambitions in that direction, but though it would make an interesting analysis of the Camphill movement, I don't think very many of the co-workers today would regard themselves as such. However, in the case of children, Camphill is closer to a monastery than to kibbutz. If you take a detailed map of any Camphill village, you will seldom find a specific area set aside for children, sometimes a kindergarten, or a school, but not often. If you look at the accounts of a Camphill village, you will rarely find children as a post in their budget, though here at Solborg we have now introduced a budget for children's work during the summer. When I studied the organizational structure of the Norwegian Camphill villages, I did not find a single committee or mandate with responsibility for children. Compare this situation with a kibbutz, which has children's houses, and often several committees, with budgets, that deal with children in various ways.

Reading through what I just wrote it might seem that this is a negative criticism of Camphill. It is not meant as such. Children are not neglected in Camphill. They are part of the strong family structure, and responsibility for the children lies with their parents. The children grow up in a safe and secure environment, they have freedom and space, and many opportunities and resources for learning. They are exposed to many different kinds of people. Growing up in a large Camphill village must be a very similar experience to growing up on kibbutz. Many of the children stay on, and form a second or third generation. Many move on to related vocations such as Steiner schools or various forms of social work. In my experience of children from Camphill and from kibbutz, and indeed from large intentional communities in general, I have found them to be better equipped to deal with the world than children growing up in the rest of society. This is not a scientifically proven statement, but a gut feeling, an intuitive impression. Both on the kibbutz and in Camphill, we have many volunteers coming through to work in our villages for shorter periods of time. Some of these volunteers come from other communities, while most of course come from the larger western society. Those that have a background of community tend to be more capable of working in a variety of practical jobs, they tend to be able to take responsibility, and they tend to get on well with others socially, working in a team. While living on the kibbutz, we had a number of young people from Camphill villages as volunteers, and they were highly regarded. In fact it was enough to mention that you came from Camphill to get a positive answer to your request to come as a volunteer. Some of them have even had their tickets paid for to come back a second or third time.

With new generations growing up within a community, the time eventually comes when they take over the running of its affairs. For most communities this handing on takes place much more often than the traditional quarter century or so needed for children to grow up to be adults. Founders get displaced, as we saw when we looked at the transition from pioneering to maturity. The managers take over from the founders, and a certain amount of structure gets put into place. When we get around to the children of the founders becoming old enough to take on the running of the show, we are already a long way down the line, and management has long since passed out of the hands of the original founders. When we talk of generations, it often means the next group, rather than the biological children of the founders.

Ed Schein comments on the handing over from one generation to the next:

> The key issue for culture change leaders is that they must become marginal in their own culture sufficiently to recognize what may be its strengths worth preserving and its maladaptive assumptions requiring change. This demands the ability to learn new ways of thinking, as a prelude to unfreezing and changing their organization. This process is especially difficult for entrepreneurial founders because the early success of their organization is likely to make them believe that their own assumptions are ultimately the correct ones.

Here he is writing not about biological generations, but about the reigns of power passing from the founders to the next group: the change from pioneering to maturity. But the inability of the older generation to let go is a factor which is always a risk whenever power is handed on. How succession is managed then becomes a major issue. Ed comments that in extreme cases a founder may unconsciously destroy the organization in order to prove how indispensable he is. He suggests that a clear cultural identity can help an organization through such processes.

Whenever an intentional community comes to handling generational succession, this is a crucial issue, and if it isn't handled right, can mean a quick transition from stable maturity to terminal old age.

17. Dynasty!

As generations grow up, even in an alternative eco-community, one might expect habits that have been acquired over thousands of years to reassert themselves, however much the intent may be to create new habits. The strengthening of family ties, combined with the tendency for some individuals to be more powerful, commanding, or willing to take on leadership roles, will inevitably make for the emergence of dynasties, extended families or groups of families. These can create connections between themselves, and collectively dominate the power structures of the community or movement. I see this tendency in kibbutz, in Camphill and in the Bruderhof. I have no doubt that the same phenomenon exists in other communal movements that have a multi generation track record.

My first reaction to this tendency was negative. I thought of nepotism, of old boys' networks, the Mafia, and of what in Hebrew is called *proteksia,* favouritism amongst groups of friends or members of a family. I thought this A Bad Thing, running contrary to my notions of equality and democracy, which were of course Good Things. Upon reflection, it may not be better or worse than many other characteristics. It has become clear to me that some individuals possess or acquire considerable skills in leadership, in focussing a group of people around an idea, and of being able to facilitate a group to arrive at decisions or solve conflicts. Likewise, other individuals have traits that cause conflicts, and that create disturbances or confusion within a group. We have all experienced good managers and bad managers, in community and in other organizations. Given that leadership people are not corrupted, not using their power for self-seeking, and not destroying those they don't like, I'm happy to encourage good leaders to lead and poor leaders to find other useful occupations, or to retrain. Should they happen to marry other good leaders, or the sons, daughters or nephews of good leaders, that in itself is not a reason to accuse them of nepotism or other skulduggery.

Maybe we should be happy that certain families produce members who can take the lead on issues, that can consolidate a group, and be able to hold the strands of a small community

From kibbutz to eco-community

Twin Oaks began as a small radical commune in the 1960s, inspired by modern psychological ideas of behaviourism. The basic aim then was to create the conditions by which a better type of human being would be nurtured, and this in turn would help solve the problems of the world. Solid universal aims.

One of the ways to create the right conditions was to go the path of collective child rearing. Looking for models they found the kibbutz to be the closest thing to what they aspired to. Their first purpose-built child rearing house was named Degania (see below), after the first kibbutz founded in 1911 near the Sea of Galilee. The people trained and entrusted to rear the children were called *metas,* after the Hebrew word for child-minder, *metapelet.* When I visited in 2004, I found this a fascinating example of how community ideas travel to places never dreamed of by those who instigate them. It was especially significant visiting with my family; Ruth had worked for years in children's houses as a *metapelet,* and our children had grown up in the kibbutz. We all knew what Degania was.

Even more fascinating was that Twin Oaks travelled the path from collective to family-based child rearing in just a couple of decades, and Degania, the building, was now being used as a family house instead of a children's house.

If it's not a law of human nature to move slowly from the more collective to the less collective, it certainly is a trend to be found in many community situations.

together. We may want to keep things on an even keel by having a system of checks and balances built in. If this consolidation of dynastic power makes it difficult for others to penetrate or be accepted into managerial roles then it certainly will be detrimental to the future of the community as a vehicle for democratic or egalitarian ideals. Our challenge is not to focus and get blinded by the family ties, but to look at the individual. Is she or he capable of doing that particular job well? Trustworthy? Honest? These are the qualities we are looking for and those are the kind of questions we need to ask ourselves.

The role of the family has been the subject of a great deal of experiment and research within the context of intentional community. From the nineteenth century onwards, many communities have had ideals of breaking down the conventional family relationships in order to create the 'better human being.' Kibbutz in the early days had that as one of its ideals. Twin Oaks, a small but very strong community set up in the mid-1960s, was inspired by B.F. Skinner's behaviourism. Many libertarians and anarchists, writing in the early 1970s regarded the nuclear family with horror, as one of the causes of human suffering and of the downfall of society.

For many decades there has been a popular view that life in community is one long free sex orgy, and that kids have no idea who their parents are. The mass media in the late 1960s and throughout the hippy era of the early 1970s found that to be great copy which sold magazines and newspapers like hot cakes. Communes became associated with sex and drug-crazed hippies, and still hasn't recovered from this unfortunate smear campaign. This had become such an accepted truism that in 2002 Veronika van Duin opens her autobiography: 'Growing up in community does not necessarily mean that one doesn't know one's parents. I knew mine very well.'

She goes on to describe her childhood and adolescence in the Camphill villages in Great Britain in the 1950s and 60s. Her mother had been one of the original group from Vienna that founded Camphill in Scotland in 1939. Veronika grew up squarely within the community, and her description of community life is one that I really recommend to anyone who wants to know what it's like from the inside. For her, the Camphill community life was natural, that's how the world was. Today, she is one of the leading lights in teaching and inspiring Camphill members in creating the homely atmosphere that is one of the most important characteristics of the Camphill tradition and culture.

Clearly, any community or movement which makes it far enough down the line to have multiple generations and weave dynastic relationships has established itself, has certainly achieved maturity, and has all the potential for arriving at old age. However, I don't feel discouraged if the intentional community, originally created as an alternative to mainstream society, creates it's own more or less comfortable social scene. Deep down, despite all the attractions of city life, of the excitement of Times Square and the hot spots of Soho and the fun of Notting Hill, I really believe that village life is what is best for the majority of human beings. I was happy to read the ending of Richard Critchfield's book on villages. He was in no way a radical alternative type, there are many opinions in his book I would take issue with, but there was a final note that I heartily agree with him on:

> History suggests that there may be no adequate substitute for this universal village culture, which, for all its old restraints, religious conventions and patterns of obedience, seems necessary in small communities of people living off the land. It just could be the most harmonious way of life for human beings who choose to live in groups.

Could it be that the urban experiment is showing itself to be inhuman? Are our ideals of creating ecovillages a way of saying that enough is enough, we tried mass urban living with extreme industrialization, and found that it doesn't work so well?

Maybe we have come full circle.

We grew up in villages for many thousands of years; we found that they were often small-minded and oppressive. Our desire for freedom grew even as our consciousness of ourselves as individuals did. We created cities, our freedom increased, our choices multiplied, but so did our egos and our capacity to exploit and oppress each other. Within mainstream society we created ever smaller units, down to the smallest family unit possible: the single parent family. We created vast numbers of alienated individuals.

Within western society we then trained thousands of psychiatrists, psychologists and social workers to help these alienated and dysfunctional individualists. Finally we are coming together as free individuals to create a new society, a new kind of dynasty, where it is the ideals and aspirations that are the glue that binds us, rather than our blood ties.

18. Old Age

The Amana society traces its origins directly back nearly three hundred years to the Pietist movement, specifically to a meeting between Johann Friedrich Rock and Eberhard Ludwig Gruber in 1714. This was a religious movement which awakened more and more people, and over the next century established itself in areas of central Europe which are now Germany, Switzerland and the Czech Republic. Being inspired by the word of God, and receiving this inspiration directly, this particular group called themselves Inspirationists and set up congregations not unlike other Christian churches. In the 1820s they experienced increasing oppression, at the same time as Christian Metz took over the leadership. He was clearly a highly charismatic personality, and descriptions of him record a humble, popular, and wise person, with extraordinary skills in management and leadership. This combination of outside pressure and inspired leadership was to prove crucial for the development of the Inspirationists from a series of congregations into one of the most successful communal movements of modern times.

As the oppression grew heavier the Inspirationist congregations gathered into estates for mutual protection, buying old monasteries and large properties. At the same time many oppressed people in Europe were attracted to emigrating to new land in America. In the 1840s, for example, hundreds of thousands of Irish sailed to the New World to escape from the potato famine. The Inspirationists had already found that by organizing themselves into communities, they could weather the storms of oppression much easier, and agreed that they would move to America as communities, rather than singly or as families. In order to pay for the voyage, and buy land in New York State, Metz instituted a constitution that looked very much like communism, inspired not by the emerging socialist and communist ideas then prevalent, but by the Biblical injunction in Acts 2:44–45 to hold 'all things in common.'

The new site, near Buffalo, was called by them Ebenezer, and in 1843 over eight hundred of the Inspirationists settled there. Within a decade they had built four villages, and acquired another two in neighbouring Canada. But there were signs that the strict religious life they had instituted was beginning to be tempted away by the otherwise irreligious life in the more easy going villages and towns of their neighbours. A search was made for land out west, and the whole community moved to Iowa in the 1850s. The site they moved to they called Amana, and within a few years they had established six villages in a large 26,000 acre tract of land along the Iowa River.

In practical terms, the difference between the Amana colonies and communism would be hard to find. They held land, machinery and tools in common. They ate in communal kitchens, and had no personal income, but received some pocket money when it was needed. Children were looked after communally, and schooling was free. But what held them together was not a political ideology, but a strong religious faith. They worshipped together, read the Bible together, and carried on being led by inspired leaders, who received their inspiration through being 'moved by the spirit.' For the following eighty years this way of life established itself, and their business acumen and hard work created a highly successful community, with no real financial hardships, and generally good relations with their neighbours.

Towards the end of this time, during the 1920s, many of the original integrated aspects of their lives began to fall apart. Members did not work hard enough, and they became increasingly dependent upon hired labour. Individuals began making personal money by making and selling things; the younger people were less observant in their religious practices. Some even played baseball! Partly for internal reasons such as these, and in the increasingly difficult economic climate after the crash of 1929, the business side of Amana was in danger of becoming bankrupt. Drastic measures were called for.

A committee was set up, proposals were made, and a vote was taken amongst all the members. The question was whether to disband the communal economic life. Overwhelmingly the vote was to do so, and within a few weeks, in the spring of 1932, Amana was reorganized into a private, capitalist, free enterprise society.

However, this is not the end of the story. A visit to Amana today is a step into something which many of us dream about. The six villages are still there, the land around is green and well cared for, the landscape is rolling and pleasant, with plenty of walks along the river, through the meadows and short distances from village to village. The houses are well-built and sturdy, and the gardens well tended and full of flowers, at least in the high summer, when I was there.

Dig a little deeper and you will find many of the original family names still current, the older people have a slight mid-European accent, and the village culture is qualitatively different from other places. There is a strong feeling of community, an identity, a shared history, and a high degree of social security. The communistic past is over. The strict communal life has been gathered together into a museum, and has become an object of research. In one sense it is dead, but it remains in the spaces between the houses. It pervades the six villages.

Thousands of tourists flock there every year to experience that feeling. It has become somewhat of a theme park experience, and there is the ever present danger that it will be trivialized, but at least the visitors are getting some shadow of an experience of a better society, rather than some fantasy of a series of cartoon characters.

The popularity of Amana is no doubt the attraction of experiencing that close-knit traditional village life I described in the Introduction, which had been so well documented by Richard Critchfield. Modern life, even in the Mid West of the USA, is fast paced, confusing and unstable. How attractive to be able to step back even for a short while into a nostalgic village past where everyone knew each other, depended upon each other and helped each other.

There were many such attempts at community during that century in North America, and they give us an opportunity to compare different systems of governance and community building. Not only that, we can also compare their spiritual background. Many were highly religious, while others were strictly secular. Mark Holloway, in *Heavens on Earth,* comments on these communes:

> Communities ... produced a high standard of living and workmanship, were pioneers in Negro and feminine emancipation, in democratic government, in eugenics, in the primitive psychoanalysis of mutual criticism, and in education and social reform. They were a benefit to their neighbours and also to the nation; and they showed by example that associative

Where have all the heroes gone?

I was very inspired when I got the opportunity to visit Amana in 2004. I feel I need to include this picture of the cemetery from Amana village in this book. The simplicity of the gravestones and the humility of the cemetery itself made a deep impression upon me. Here lay the remains of hundreds of community members who had created a window through which we could glimpse the vision of a better society. Not by preaching and writing or being intellectual, but by living their vision into real life. By the way they worked together, ate together, brought up their children together, and worshipped together.

I was filled with a strange feeling as I walked around looking at the old buildings, as I talked to people who lived there, and as I read their history. It comes back to me again when I pick up the excellent little book about Amana by Peter Hoehnle that I brought back home with me. This feeling seems to crystallize itself into thinking nostalgically how good life must have been in Amana during its heyday. A community working together, not having to worry about individual wages and bills, but really able to put their best energy into the task at hand: building a house; growing crops; cooking for large groups and bringing up children; the sense of security; the slow walk to services on a Sunday morning, where instead of being served hell-fire sermons by priests, they talked quietly amongst themselves how they could live lives closer to the example given by Jesus.

I know there are no communities without problems and conflicts, but I have had glimpses of what it might be like if we can avoid them or solve them for a while. Amana was one of those glimpses.

> effort of this type can be highly satisfactory. It is well for us that these examples exist, with all their faults and follies; for whether we like it or not, it is always possible that necessity may force such a life upon us. If so, we should be grateful for this fund of experience.

I'm not quite sure what eugenics is doing in there, but taken as a whole, it's clear that even if the communities themselves have disbanded or faded into the main fabric of society, they have contributed many positive features to that society.

Old age in the context of community development may look like the dissipation of collective ideals, and some commentators conclude that there is an inevitability in the demise of the co-operative dream. It may well seem so, that however hard we try to create an utopian ideal here on earth, it always falls short, and the vision fades away with time, being replaced by good old egoism and self-seeking.

> The first generation has an idea and lives for that idea. The second generation perpetuates that idea for the sake of their fathers, but their hearts are not in it. The third generation openly rebels against the task of mere perpetuation of institutions founded by their grandfathers — it is always the same with people. (F. William Miller, Amana pharmacist, 1933. Quoted in Peter Hoehnle's *The Amana People.)*

There used to be a saying amongst the British upper middle classes in the nineteenth century that a family creates and destroys itself in three generations. The first builds up a successful business or property, the second inherits this and carries it along, the third squanders the wealth on good living. Whether this can be applied to intentional community I can't really tell, but the idea is worth reflecting upon. Certainly the Amana pharmacist thought so. Relatively few community movements seem to get beyond the third or fourth generation without slowly becoming indistinguishable from the surrounding society, even though they may retain some exotic cultural features; songs, styles of furniture making, dress, funny accents.

But look again. Check it out in the context of the surrounding society, and look closely at how that society gradually changes. Chris Coates listed some contributions that intentional community had given to British society and ends his review of British utopian communal experiments by asserting the following:

— The ideas of the Italian educationalist Heinrich Pestalozzi, were taken up by the Sacred Socialists and by the Owenites in the early nineteenth century, and now form the basis for the entire British educational system.

— Modern clothes, introduced by the Healthy and Artistic Dress union, were pioneered as 'Rational Dress' throughout utopian communities around the turn of the twentieth century. Up to that great loosening of fashion in the late 1960s and early 1970s, that's what we were wearing in Western Europe.

— The entire British Town and Country Planning System was invented by Ebenezer Howard in his Garden City idea, and developed by the arts and crafts movement.

— Four British Prime ministers have strong connections to utopian communities: Ramsey Macdonald resided at the Fellowship of the New Life, Herbert Asquith was educated by the Moravian Brethren, Clement Attlee was secretary of Toynbee Hall and Harold Wilson lived in Hampstead Garden Suburb.

To quote Chris Coates directly:

> And finally perhaps the biggest utopian experiment of them all — the Welfare State. Penned by William Beveridge, former secretary at Toynbee Hall and editor of the Ruskinite paper 'St. George,' it was in essence an attempt to distil the experiences of two hundred years of small scale utopian experiment into a grand practical plan to deliver utopia to the masses.

Coates does not mince his words:

> The communities, as with most other utopian experiments, can be seen as experiential, social and spiritual research and development departments for society as a whole — providing a supportive environment for the pushing of personal and social boundaries.

As old age takes over, and these intentional utopian communities dissolve into the mass society, what they are achieving is a change in that very society that they originally set themselves up against. The communities themselves may dissolve, the buildings are taken over and fulfil other functions, the members die off, drift away, or retreat into nuclear families. But the ideas remain. Not only remain, but often subtly seep into the mass society, and become part of the accepted norm.

Don Pitzer suggests that ideas are introduced into the social realm through community and live on after the community phase of the idea is over. Communalism, in his view, is merely an initial 'ephemeral' period in the life of an idea. As examples he quotes Christianity, Seventh Day Baptists, Mormons, Robert Owen, Fourier, all of whom introduced ideas that went through a communal phase which faded away, leaving the ideas to establish themselves throughout society. He suggests that communal living is a vehicle rather than a destination. The whole idea of measuring the success or failure of community should be directed at how much the idea behind the communal impulse goes on to penetrate the wider society (you can read more about Pitzer's theory of 'Communal Developmentalism' in a fascinating article in the 1993 edition of *Diggers and Dreamers).*

One of the images which I have had with me on this journey through the life cycles of community is that of the river. In youth the river is lively, noisy and vigorous as it rushes down the mountainside. In maturity it is grand and stately, winding through rolling countryside, watering the meadows and supporting great trees along its banks. In old age the river creates deltas and finally empties itself into the sea, losing itself in the ocean. Without rivers, though, the ocean would dry up through evaporation, which is what is happening with the Dead Sea, the Caspian and the Aral Sea.

Rivers keep the ocean alive.

The alternative becomes the establishment

The community of Woodcrest Bruderhof in Rifton, New York, is a youthful fifty years old.

The first time I met members of the Bruderhof movement was at a community conference in Israel in 1985. There were hundreds of people from communities worldwide. The kibbutz movement was then over seventy years old, and most of the kibbutz participants thought of themselves as representing an old and wise movement, able to give advice to the younger community experiments represented there. The Bruderhof delegation stood out clearly, the men had beards, clumpy work-boots and checked shirts, and braces (suspenders to you Americans) held up their blue trousers. The women wore simple flowered ankle-length

frocks, and their hair was tied back under their spotted headscarves.

At a keynote panel debate, one of the Bruderhof elders stood up and said that he was so happy to meet this young kibbutz movement, which had got through their first seventy years so well. He then went on to explain how the Bruderhof looked back four hundred years to their own origins, in the communities of the Moravian Bohemian Brethren founded in Eastern Europe during the sixteenth century. These developed into the Hutterites, a long and complicated history of persecutions, stability, emigration to North America and migration to the Mid West and the Great Plains. During the 1920s a group of young German and Austrian intellectuals was so inspired by a visit to the North American Hutterites that they set up their own Bruderhof communities in Germany as part of the Hutterite church. When the Nazis came to power they moved to England. As the Second World War broke out, it became unfortunate to be German in England and they moved to South America. After two decades it was apparent to them that life was just too hard there, and they moved to North America in the 1950s, and so it was that we arrived to visit Woodcrest and found their fifty year birthday decorations still in place.

The Bruderhof are one of the few examples I know of a community which has not moved significantly towards individualization. They are family-based and always were, but still eat most of their meals in large dining halls, don't have personal wages, and make their decisions at large general meetings. The only signs of degeneration that I could see were that many of the men had stopped wearing braces and checked shirts. Some had begun to wear jeans!

Is this the beginning of the end for the Bruderhof?

In a similar way, intentional community keeps society alive, it contributes social impulses, and social renewal. They say that a great river such as the Amazon can be detected over two hundred miles out to sea by the colour of the water. In a similar way great communities may be detected many years later by the social changes they create. Bearing this in mind, and looking at the situation today, and especially at the vigorous development of ecovillages, we may experience many of the trends I have traced, from youthful characteristics through the mature phase, and even the gradual dissolution of old age, to be hopeful signs of social renewal rather than depressing symptoms of decrepitude.

Jonathan Dawson certainly hopes so in his recently published book *Ecovillages:*

> This is a moment of opportunity for ecovillages. That opportunity is to dare to leave the safe niche of 'being alternative,' and to enthusiastically embrace the challenge of helping mainstream society over the next several decades.

Afterword: Does a Group Have a Soul?

If you don't believe in anything beyond the purely material world you're not going to get much out of this Afterword. But if you've managed to hang in here up to now, you'll probably want to see if all this has a happy ending.

Having traced how community seems to have a life of its own, how it resembles an organism in its development through various phases, the question arises if indeed there is a spiritual component below the surface, guiding and informing the group.

This Afterword is based on the idea that when a group of people come together, something arises in the space between them which has a non-material life of its own. I would call this the group soul or the group spirit.

Given that we accept that there is a spiritual component behind, under or over the physical manifestation, we have to ask ourselves what happens to this spiritual component when the physical part dissolves. Is it subject to the same physical laws of growth, decay and death? Does it live on after the body is gone? These are pretty big questions, and we need to tread lightly for fear of offending different religious preferences. But we can try to keep an open mind, suspend dogmatism, and ask, what if?

What if there is a soul component in community? How might it behave? What does it do when the community disbands? Where did it come from in the first place? If these questions offend you, trample on your religious or non-religious convictions, or upset your view of the world, then don't read on. I have no intention here to offend anyone, or ram my ideas down anyone's throat. But if you are fascinated by the phenomenon of intentional community as a living agent of change in our western world, let's explore some of the more far out possibilities of how community behaves and why. Taste it and see; if you don't like it, you can always spit it out and go back to conventional foods.

Arnold Mindell asserts that 'community is dreaming together' which is a good starting point for putting community into a spiritual context. Now there may well be those who assert that dreams are just neurons zipping around in an over fertile brain, but for me, my dreams are real, vivid experiences. Whenever I do remember them, I seem to have come back from some fantastic other world, peopled by folks I seem to know, but who do many crazy things. I have other dreams too, daydreams of a better world. Sharing these daydreams with others is certainly what we do in intentional community and ecovillages.

Any human group will develop a greater or lesser degree of togetherness. For those sensitive to this, this can be experienced as a spiritual entity, and some communities actively cultivate this. In the development of human consciousness, we are coming towards the end of an era of intense materialism, and I feel that much of the conflict in today's world is tied up with the tension between materialism and spirituality.

It has been asserted that we cannot create a new world by using materialist thought processes alone, something else is needed. Many of today's intentional communities are exploring spiritual means of solving problems. Internal problems and conflicts are often due to misunderstandings, poor communications and differences in style and strategy rather than fundamental disagreements. External problems are greater, they can be as fundamental as creating a whole new world order. Certainly that would be in a good community tradition. It is an ambition that can be heard again and again throughout the history of intentional community.

Margarete van den Brink sees this ambition of creating an improved larger whole as the final stage in transforming people and organizations:

> In Phase 7, the organization aims in particular at the contribution it wants to make to the development of the greater whole. Here again we see that spiritual development in people always continues. While the emphasis on the development of awareness lay on the personal in phase 5 and on relationships with others in phase 6, we see that awareness and effort of organizations in phase 7 are aimed at care for, and further development of, the greater whole of which we are part: society, mankind, nature, the earth, the world, the cosmos.

Community is between people

Community arises when people get together. It somehow arises in that space between them, and gradually takes on a life of its own. It feeds off those who form community, their positive inputs, their shared dreams and collective tasks.

Once the initial work has been done, and this community entity has begun to function, things begin to happen that can't always be explained by the purely rational or purely material. We can of course call them coincidences, chance happenings. Stories abound of communities reaching some crisis point and then the solution turns up one way or another. In our community at Solborg a house came up for sale immediately adjacent to our village, and we felt that we had to buy it, because it would give us an opportunity to expand the village. As we were buying the house we had no concrete plan for who was going to live in it. A week later someone called up from another Camphill village looking for somewhere to live close to us. The building was filled immediately.

A similar thing happened a few years previously, when a couple returned from a free year of travelling, and we didn't know where to house them upon their return. Another house came up for sale next to the village, we snapped it up, and they got a great place to live.

These are things that happen, every community can tell you lots of these stories. They become part of the shared history, the myths, the culture of the community.

These are not small ambitions, and go far beyond the nuts and bolts planning that were needed in order to set up ecovillages to begin with. At that starting point our energy was concentrated on organic farming, on building with ecologically friendly materials, and making sure that sewage treatment was biologically responsible. Now, we are dealing with spiritual matters on a truly cosmic scale. Now we need to trust in other guides, using other instruction manuals. That's what Diana Leafe Christian in *Creating a Life Together* is writing about:

> Trust that it's meant to be, that you're being guided by a higher power. Diana Brause: 'After so many synchronistic events that didn't fit the scientific odds, I chose to act as if some higher force was really in charge, that the project was really a kind of sacred trust that we were privileged to take on ... that things were actually being taken care of.'

Eberhard Arnold from the Bruderhof doesn't hide his religious convictions in any way. In a short but intense book, he and Thomas Merton address this question of why they live in community. I would recommend it to anyone interested in these ideas. Eberhard has a wealth of communal living experience to draw from, and answering his own question 'Why we live in community?' he goes on to say: 'We must live in community because all life created by God exists in a communal order and works towards community.'

Even though he doesn't mention the river and the tree, which have been my own images while writing this book, they too belong to the order of all life created by God. His assertion that communal living is following a pattern laid down by God can be made acceptable to those who don't want to be so specific about their religion. The patterns we see in nature show us that all life is a community, nature operates out of community, and mutual interdependence and the interlinking of all life must encourage us to do likewise.

Eberhard goes even further, and asserts that: '... the realization of true community, the actual building up of a communal life, is impossible without faith in a higher power.'

Lucky for those who don't want to be too specific. Most of us can feel pretty comfortable with 'a higher power.' Thomas Merton, commenting on what Eberhard Arnold wrote, says: 'What he wants to stress is the fact that community is not built by man; it is built by God. It is God's work, and the basis of community is not just sociability, but faith.'

The idea that the spiritual world is working through us, encouraging, helping, and inspiring us to carry out good works in the world has been around for a long time. I was recently sent a quote from W.H. Murray's *The Scottish Himalaya Expedition,* (1951):

> Concerning all acts of initiative (and creation), there is one elementary truth the ignorance of which kills countless ideas and splendid plans: that the moment one definitely commits oneself, the providence moves too. A whole stream of events issues from the decision, raising in one's favour all manner of unforeseen incidents, meetings and material assistance, which no man could have dreamed would have come his way. I learned a deep respect for one of Goethe's couplets:
>
> > Whatever you can do or dream you can, begin it.
> > Boldness has genius, power and magic in it!'

The couplet is taken from a translation of Johann Wolfgang von Goethe's play *Faust.* It's a great quote, and the person who sent it to me was using it as an inspiration in the setting up of a new community.

The Austrian thinker Rudolf Steiner was tremendously influenced by Goethe; in fact his first job was working with Goethe's archives in Weimar. Steiner had his own thoughts about community, and how the evolution of human consciousness had moved from community based on blood ties and tradition, through the fragmentation of pursuing individual freedom, and back towards community again, this time out of free choice.

In the book *Rudolf Steiner Economist* by Christopher Houghton Budd, I found the following quote from Steiner:

Community lives on

The original farm where the first kibbutz Degania was founded, now a museum, lies a little distance from today's kibbutz of the same name. The actual year of founding is disputed, being quoted as 1910 in some books, and 1911 in others.

In the story of intentional community, the kibbutz experience will stand out as a remarkable achievement, and one that is packed full of useful and valuable experience. There has been a good deal of criticism of kibbutz, some of it related to the general political situation in the Middle East, and the behaviour of the State of Israel towards the Palestinians. It would take another and completely different book to go into that. The fact remains that kibbutz has a lot to teach us.

This picture (see opposite) was taken during a post-conference tour in 1995. This was the regular International Communal Studies Association conference, with about 250 delegates, and a busload of them went to visit sites in Northern Israel for three days. One of the stops was Degania. There must have been at least fifteen nations represented, a mixture of academics and activists. And here we all were, hearing about the founding of the first kibbutz, about their ideals and aspirations, about how they took these and translated them into everyday life.

Whatever happens to Degania afterwards, this is a supreme example of how the idea lives on, how it is cared for by those who record and preserve the stories of the past, and hand those stories on to new listeners. Who knows what effects this short visit had? Which of these activists heard a story or an idea, and brought it back to their home community, where it was heard again by those who today are struggling with the challenge of creating community? Which of these academics heard a fact that later would emerge in some article or lecture on the inner workings of communal life?

There's nothing mystical here, it's just how stories get handed on, and carried through generation after generation. But as a final comment to the pictures in this book, I would like to say that this is for me the real meaning of sustainability. The ability of our ideals to live on, to inspire others, to give meaning and to recreate society.

It is in the end the sustainability of our ideals that will create sustainable eco-communities.

> The evolution of individuality is such that the human being passes out of a condition in which he is part of the community, subservient to it and dependent on it, through the processes whereby the ego emancipates itself from the community, and thence into a condition in which the independent ego either goes off on its own, or recreates human community out of its own.

In *The Inner Aspect of the Social Question,* Steiner is even more specific:

> This is an important characteristic of spiritual life: it has its springs in freedom, in the individual initiative of the single human being, and yet it draws men together, and forms communities out of what they have in common.

Rudolf Steiner is the inspiration behind the Camphill movement, and he has inspired many other social reforms, and attempts at setting up community, including the Sekem community in Egypt (see Chapter 10). Its founder, Ibrahim Abouleish came from a well-off and devout Egyptian Moslem family. He studied in Austria and lived in Europe for many years, during which time he read Steiner's books and studied his philosophical and practical approach, known as anthroposophy. He then moved back to Egypt and created the community of Sekem, literally building up from nothing in the desert. Sekem's principles were based on a mixture of Islam and anthroposophy, a truly spiritual community. In Abouleish's own words:

> Since founding Sekem I have worked with questions relating to which circumstances give rise to creatively shaped and living forms, and how they remain dynamic. The help of the spiritual world is required to achieve a living form, which is revealed to people who are open to inspiration through their own spiritual work. Because of this there has been a continuous spiritual work among the Sekem employees from the start. This ongoing spiritual work radiates into all areas of our dealings and creates the solid foundation for the future development of Sekem.

In Camphill, this is clearly understood, and membership of the Camphill movement is dependent upon something quite subtle: whether the community has members of Camphill Community. This is a group which has a clear membership process, regular local, regional and international meetings, but very few archives, no real statutes as such, and no financial or legal existence. To quote from a letter written after a recent international meeting: 'We do not want to give a report as this could fix what is living and fluid.' Members of Camphill Community keep the spiritual aspects of community alive by being aware of it, studying it, and letting it permeate the outer, everyday work they carry out in their community.

Communities are at heart spiritual creations. They are held together by a web of relationships that spring from the spirit. The material forms, the buildings, the fields, the technology and the economy are all dependent upon these subtle relationships between the individuals that make up that community. If these relationships fall apart the community will also fall apart. We human beings are like that too. We are kept alive by a spiritual identity, our soul. With modern medical technology we can keep the body alive artificially, we can hook up lots of machines and keep the heart pumping and the lungs breathing. But if the spirit passes out of a person, he or she will die. The nails and the hair will continue to grow for a few hours, out of sheer momentum, but one organ after another will cease to function.

We can keep community artificially alive too, the buildings can still stand, the fields still be cultivated, and people can seem to go about their daily tasks. But without that subtle web of relationships that builds community, the spirit of community will cease to function.

I want to end this Afterword with a story. I first heard it from a storyteller in Israel who claims that it is an old Hassidic story. I've seen it performed as a play here in Camphill, and I read it once in a collection of stories published by the Bruderhof. Someone told me Tolstoy includes it in one of his collections of short stories. Wherever it comes from, it's a wonderful story, and I've used it several times in community contexts. This is my own version of the story. It says more about group soul than any amount of theory or quotes from other people.

A Story about Community: The Monastery That Was Saved

Once upon a time there was a fine monastery, large and well known. People from all over the world flocked to visit it and pray there. But as the years went by, monks died, there were fewer new novices, and in the end there were just five old monks left. And very few visitors.

'What shall we do?' They asked each other.

They had no idea, but one of them remembered that there lived a wise Rabbi deep in the forest, and maybe they could go to him and ask for advice. One of them went; it took him a long time to find the Rabbi. He asked him what they should do.

The Rabbi didn't know anything about monastery life. He had no advice, but he invited the monk to share his life for a few days before returning home to the monastery. It was a quiet life, filled with prayer and simple work.

At last the monk took his leave, and then the Rabbi told him that he had had a dream about the monastery that had made no sense to him. The Rabbi had dreamt that one of the monks was the Messiah. Surely it was just a silly dream with no meaning.

The monk returned home, and the others asked eagerly if he had been given some good advice. He had nothing to tell apart from that silly story about the dream. About how one of them was the Messiah.

But they began to wonder. Could it be Brother Paul? But then he was always so short tempered and angry. Could it be Brother Matthew? But he was so simple. Could it be Brother Augustine? But all he did was pray, and never helped with the practical work.

But something changed in the way they behaved towards each other.

Visitors still came from time to time, and they noticed the fine atmosphere in the monastery, how the monks respected each other and behaved courteously towards each other. The visitors spoke about it after they went home, and more visitors arrived. One day a novice arrived to ask if he could stay.

The monastery began to grow, its fame spread again far and wide. Pilgrims flocked to the place to experience the very special atmosphere that the monks had created amongst themselves.

References

I must begin by mentioning two books which were basic to preparing the present work. These two books were consulted so often that it seemed to me that they warrant a special mention.

Christensen, Karen and Levinson, David. *Encyclopaedia of Community. 2003.* Sage Publications, California, London and New Delhi. Four volumes.
A landmark work. Four thick volumes, written by over three hundred contributors, and containing a vast overview. Browsing the material will give you endless hours of interesting and fascinating ideas about community in its various aspects. Highly recommended that you get access to it through a library. When your community gets rich enough, order a set for your own book collection.

Bang, Jan Martin. *Ecovillages — A Practical Guide to Sustainable Communities.* (2005) Floris Books, Scotland.
Written as a planning guide for those wanting to set up an ecovillage, and based on the Permaculture Design Course, this lays the foundation for much of the material in the first part of this present book. It begins with a longer background to the ecovillage movement, and a more comprehensive description of what Permaculture consists of. I tried not to repeat parts of this when writing Part 1 of the present book, which forms a fitting sequel to the first book.

Apart from these two exceptions, the books I have referred to while writing *Growing Eco-Communities* are arranged by chapter when they are first mentioned or relevant. This means you may have to hunt through this section, looking for the details of the book you are interested in. Not such a bad thing; during your hunting, other books may catch your eye. Enjoy those random connections. To help you find a specific book quickly, I have added another list under 'Book List,' where titles are listed alphabetically by author.

Acknowledgments

Gering, Ralf. (2002) *Encyclopedia of Large Intentional Communities.*
Only available online. It's just a list of communities, networks and movements. An amazing work of research which I got from a chat group I was part of a few years ago. Email enquiries to Ralf asking for updates haven't been answered.

Foreword

Glasl, Friedrich. *The Enterprise of the Future.* (1994) Hawthorn Press, UK.
A short book of just one hundred pages, this is a treasure trove of ideas and observations of how people work together. With a basis in anthroposophy and business management, Glasl constructs a developmental process which should be of interest to anyone who thinks about how human groups develop and grow.

Introduction

Critchfield, Richard. (1981) *Villages.* Anchor Books, USA.
Written by an international expert long before ecovillages were invented as such. Taking traditional villages as a starting point for social change, I found this a fascinating study in what our base consists of. A monumental work, lots of details, and lots of great stories from around the world.

Handy, Charles. (1991) *Gods of Management.* Arrow Books, UK.
Originally published in 1978, this is a really interesting analysis of management styles, arranged and named after four Greek gods. My friends in management consultancy assure me that Charles Handy is one of the great management gurus, and there is certainly a lot to be learnt from him about how groups of people work together.

Brink, Margarete van den. (2004) *Transforming People and Organizations.* Temple Lodge, UK.
Margarete links the developmental steps that a group goes through to the same steps that an individual goes through, based on ideas formulated by Rudolf Steiner. These steps form the foundations for the highly successful Waldorf School movement, and have proved themselves there. This book was one of the first confirmations for me that my own vague ideas about community development had some relevance, and gave me tremendous inspiration for continuing my own train of thought. Highly recommended.

Communities Directory. (2005 edition) The Fellowship for Intentional Community, USA.
Though most of the entries are from North America, this directory is useful for anyone interested in the communal living scene. Lots of good articles giving background information and viewpoints, give it relevance far beyond the shores of that continent. And it's fun just

to read about all those communes that are listed, some of them you might have heard of, others are new eye openers.

Schein, Edgar. (1999) *The Corporate Culture Survival Guide.* Jossey-Bass publishers, San Francisco, USA.
Recommended to me by friends from the business management world, this was one of my first eye openers into how much we in the alternative movement can learn from the mainstream. Corporations create their own cultures, as do ecovillages and other communities. Being aware of this, and how these local cultures change over time, helps us to become active co-creators.

Foy, Nancy. (1980) *The Yin and Yang of Organizations.* William Morrow and Company, Inc. New York.
Nancy Foy has her background in mainstream management; her most famous book is about IBM. By taking the Yin/Yang model grounded in ancient eastern Taoism, she has arrived at fascinating insights into how large corporations behave, insights which are often equally relevant to those of us observing how communities behave.

Gelb, Saadia. (1994) Almost *one hundred years of togetherness.* Kfar Blum Guest House, Israel.
Saadia is a veteran kibbutz activist, with a lifetime of experience at nearly all levels of kibbutz life and organization. This little, privately printed book is a treasure trove of humorous insights into communal living, from the tragic to the ridiculous. This book will give you a feeling of what it's really like to live in community.

Leon, Dan. (1964) *The Kibbutz.* Israel Horizons, World Hashomer Hatzair, Israel.
Outdated, to be sure, but still relevant as a snapshot of kibbutz while Israel was still a young, naïve and pioneering venture. Really good detailed account of how kibbutz was organized at that time, and its relationship to the rest of Israeli society.

Saunders, Nicholas. (1975) *Alternative England and Wales.* Nicholas Saunders, UK.
This is a gem, a real collector's item. Nicholas has given us a snapshot of the alternative culture specific to England and Wales in the mid-1970s. It covers an amazing range of subjects that alternative folks were into at the time, and gives contacts and local information in extraordinary detail. This was England's equivalent to the Whole Earth Catalogue which came out of California a few years earlier. For those of you who were there then, this is nostalgia unlimited!

Bailey, Keith; Bang, Jan and Matthews, Bob. Eds. (1986) *The Collective Experience.* Communes Network, UK.
This is a collection of the best articles culled from a decade or more of back issues of Communes Magazine and newsletters from Britain. Real experiences, written by and for community activists. A collector's item!

Dawson, Jonathan. (2006) *Ecovillages— New Frontiers for Sustainability.* Green Books, UK.
The latest authoritative book on where ecovillages are now, written by the President of the Global Ecovillage Network (GEN) and Executive Secretary of GEN-Europe. Short and to the point, Jonathan's study covers the history of GEN, the present situation, and points out directions for the future. And he manages to do this in a global context, drawing on examples from all over the world.

Directory of Intentional Communities. 1991 edition. Fellowship for Intentional Community, USA.
The factual information has been outdated by more recent editions of the directory, but this issue contains well over a hundred pages of in-depth articles covering many aspects of communal living. Invaluable.

Kant, Joanita. (1990) *The Hutterite Community Cookbook.* Good Books, USA.
Mostly that, just a cookbook, but gives a good introduction to who the Hutterites are, where they come from, and what their communities are like.

Part 1. Pioneering Phase

1. Everybody Does Everything

Bock, Friedwart. Ed. (2004) *Builders of Camphill.* Floris Books, Scotland.
This is a compilation of short biographies of the members of the original group that formed in Vienna in the 1930s and moved to Scotland to found Camphill in 1939. These were not ordinary people, and a fascinating group picture builds up as you get to know each one in turn. A must for Camphillers, and highly recommended for anyone wanting to know more about the early days and growth of this network of ecological and anthroposophical villages.

Schwarz, Walter and Dorothy. (1987) *Breaking Through.* Green Books, UK.
Walter and Dorothy spent a long time travelling the world and experiencing communes of different kinds. They are both established writers, Walter having a lifetime of writing for the *Guardian* newspaper, and Dorothy teaching creative writing. So you are guaranteed well written stuff. Combine this with their experiences and insights, and you get a winner!

2. Decisions over Dinner

Ansell, Vera; Coates, Chris; Dawling, Pam; How, Jonathan; Morris, William and Wood, Andy. Eds. (1989) *Diggers and Dreamers.* Communes Network Publication, UK.
One of a series of directories documenting the British commune scene over a couple of decades. In addition to the directory, each edition has a series of in depth articles about many aspects of communal living.

Christian, Diana Leafe. (2003) *Creating a Life Together.* New Society Publishers, Canada.
Diana has edited *Communities Magazine* for many years, and observed that about 90% of aspiring community ventures never get very far. This books sets out to reduce that fall off rate. Full of advice based on observing real life scenarios, and written by someone who herself lives in community, it's highly recommended for anyone thinking about starting or joining a community.

Gorman, Clem. (1975) *People Together.* Paladin, UK.
An account of communes in England from the early 1970s. Good straightforward stuff.

Jackson, Hildur. Ed. (1999) *Creating Harmony.* Gaia Trust/Permanent Publications, UK and Denmark.
Hildur has been a key person in the ecovillage movement in Denmark, and internationally. She sees community as one way to explore and deepen our relationships to each other, and in this book she has collected the wisdom of many of the really experienced group dynamic and conflict resolution guides that are spread throughout the commune world. If you never have conflicts in your community, you don't need to read this book, just write to Hildur telling her the secret of your success. If you do experience conflict, then this book might give you some pointers towards resolving them.

5. Open the Door or Close the Gate

Eno, Sarah and Treanor, Dave. (1982) *The Collective Housing Handbook.* Laurieston Hall Publications, Scotland.
This is the collected wisdom from experience by a group of commune founders who had more than a decade of experience behind them when they wrote it. Though short, and with a lot of British specific information regarding legal models, it remains a book that I have often gone back to. The sections dealing with group dynamics and decision-making methods are timeless and relevant.

6. Flexibility and Multi-Use of Buildings

Allen, Joan de Ris. (1990) *Living Buildings.* Camphill Architects, Scotland.
Joan designed the house that I live in. One of the pioneers behind Camphill Architects, who provide a service for all Camphill communities around the world, Joan uses this book to give an account of fifty chapels and halls that play such a central role in Camphill communities. Having established a style unique to Camphill, firmly based upon anthroposophical architecture, this is a glimpse into true community culture. And it's really well illustrated!

Part 2. Maturity and Stability

Introduction

Luxford, Michael and Jane. (2003) *A Sense for Community.* Directions for Change, UK.
The result of several years of research and travel by two seasoned Camphill co-workers. Very much an insider book, it would be difficult for someone not familiar with Camphill life to understand, but for us Camphillers this is an invaluable insight into what is going on today.

Steiner, Rudolf. (1992) *Ideas for a New Europe.* Rudolf Steiner Press, UK.
Transcripts of seven lectures which Rudolf Steiner delivered in the aftermath of the First World War. He presents a wide view of history,

of the changing consciousness of human beings, of architecture and economics as reflections of developing humanity. Read this book slowly, every sentence requires deep thought.

8. Kick out the Troublemakers!

Mindell, Arnold. (1995) *Sitting in the fire.* Lao Tse Press, USA.
Arnold is the co-founder of Processwork, and has written numerous books on many aspects of conflict resolution. One of his fascinating insights is that conflict is necessary for group development, and that by dealing with conflict head on and with an open mind, we can arrive at much deeper understandings of who we are and how we behave. Highly recommended.

9. Building the Building Business

Jacobsen, Rolf. (2001) *Tun, bygninger og økologi.* (Courtyards, buildings and ecology) Landbruksforlaget, Norway.
I know most of you don't read Norwegian, so this book will be outside your grasp. But I mentioned it, so I owe you a reference. Rolf is Norway's leading ecovillage designer, with a great deal of experience. This book charts the emergence of the ecovillage idea in Norway, through the various alternative community ventures during the 1980s and 1990s. Full of drawings, plans and good, sensible advice.

10. Selling Yourself in the Wider Economy

Abouleish, Ibrahim. (2005) *Sekem.* Floris Books, Scotland.
This is an amazing story of how a young Egyptian goes to Austria to study, settles there, gets very deeply involved in anthroposophy, and as a mature man, returns to Egypt with his Austrian wife and children to establish an ecological village amongst the desert tribes. His continued commitment to Islam, combined with his anthroposophical ideals, results in an enormous enterprise growing organic cotton, and creating a business network of astounding size and significance. An inspiring tale of cross-cultural fertilization and practical social improvement. A couple of years ago Ibrahim received the Right Livelihood Award for his work, and increasingly I come across people who are building links with the core settlement, Sekem.

Groh, Trauger and McFadden, Steven. (1997) *Farms of Tomorrow Revisited.* Biodynamic Farming and Gardening Association, USA.
The book of my choice when it comes to Community Supported Agriculture. This is an expanded reprint of the first book they wrote, and is based on real life experience. Anyone wanting to get into CSA should look at this book first. Why not learn from the experience of others?

Lilipoh Magazine, 'Community Supported Anthroposophic Medicine.' Issue 43, Volume 11, Spring 2006. PO Box 628, 28 Gay Street, Phoenixville, PA 19460 (see also www.lilipoh.com).
I was really inspired when I read how these people had taken the Community Supported Agriculture (CSA) idea and applied it to the medical world. Maybe we can think of other applications?

12. The Poor Are Always with Us

Fike, Rupert. Editor. (1998) *Voices from The Farm.* The Book Publishing Company, Tennessee, USA.
In this book, members of The Farm talk about their community and how it changed over the years. These are authentic accounts of ecovillage life as it is lived. Read it and get to know one of the most interesting alternative communities of the 1970s.

13. I Want My Own Room

Bettelheim, Bruno. (1971) *The Children of the Dream.* Granada Publishing, UK.
This is a classic! Written in the early 1960s, it is now a snapshot of a lost past, but really full of insights and observations of how the classic kibbutz child-rearing system worked. A must for all behaviourists.

Part 3. Old Age

Coates, Chris. (2001) *Utopia Britannica.* Diggers and Dreamers Publications, UK.
One of the most inspiring books I have read in many years. Never before did I realize how much we owe to the early pioneers of communal living, and never before have I seen so clearly stated what

deep and far reaching effects the ideas behind communal living have penetrated into mainstream society. If anyone thinks we are marginal and ineffective, I highly recommend this book as an antidote to such blasphemous thoughts.

15. I Want My Own Money

Holloway, Mark. (1966) *Heavens on Earth.* Dover Publications, USA.
First published in 1951, the copy I have is a 1966 reprint. But it doesn't really matter, as Mark deals with American communities from 1680 till 1880, and the only things missing are the interpretations and insights that have emerged over the last half century. Read it for a short but thorough account of a commune scene that really made an impact.

Hoehnle, Peter. (2003) *The Amana People.* Penfield Books, USA.
The Amana community is a classic example of how a community movement arises, develops and fades away. The surprising thing is that even in their fading away, the shadow they cast is so strong and so long, that it remains a tangible experience to visit and appreciate their achievements. My own visit to Amana was immensely enriched by reading this book, and having the opportunity to talk to Peter while I was there.

Steiner, Rudolf. (1974) *The Inner Aspect of the Social Question.* Rudolf Steiner Press, UK.
Three lectures given by Rudolf Steiner in 1919, when he was offering solutions to the turmoil created by the First World War. Fascinating insight into the connections between politics and economics, and how these two are working on different aspects of our spirituality. Dense stuff, not for the faint hearted!

16. The Next Generation

Gaskin, Ina May. (1977) *Spiritual Midwifery.* The Book Publishing Company, USA.
Community culture strikes again! After The Farm midwives had handled over a thousand births they felt confident enough to issue this account of some of them, plus statistics showing that their track record was better than most high-tech midwifery clinics in North America. The book helped establish them as a leading teaching

team, which has slowly had an enormous effect upon midwifery in general. At a personal level, Ruth and I found this book's blend of common sense and spirituality of tremendous help in the birth of our own three children.

17. Dynasty!

Van Diun, Veronika. (2002) *A Child in Community.* Upfront Publishing, UK.
If you've ever wondered what it's like to grow up in the world of community, read this book. Well written, observant, thoughtful and at times funny. Veronika grew up in Camphill, and this is Camphill seen from the inside!

18. Old Age

Pitzer, Don. (1993) 'Developmental communalism.' Article in *Diggers and Dreamers.* Communes Network, UK.
Don is one of the great thinkers of the communal world. His ideas about how community fits into the wider social context are tremendously stimulating. I owe him a great debt.

Afterword: Does a Group Have a Soul?

Budd, Christopher Houghton, and Steiner, Rudolf. (1996) *Rudolf Steiner Economist.* New Economy Publications, UK.
It's generally thought that Rudolf Steiner lectured and wrote only about esoteric and spiritual subjects, so it's a surprise for many to find him seriously engaged in economics. The fact is that not only did he give practical advice to business and politics, but many of his ideas have been put into practice subsequently. This little book relates Steiner's ideas to the 'New Economics' movement, and is a real eye opener!

Merton, Thomas. (1960) *The Wisdom of the Desert.* Shambhala Publications, USA.
A tiny little book that I sometimes take with me on lightweight trips. Full of quotes from the earliest Christian hermits and monks who went out into the Egyptian desert to get away from it all. They gradually got together in monasteries, creating a tradition of monasticism which carries on to this day. These insights into living together in community are still as fresh and as relevant as ever.

Book List

These are the same books as listed above in 'References,' but here put in alphabetical order, arranged by author. This may help you in navigating for a specific book.

Abouleish, Ibrahim. (2005) *Sekem*. Floris Books, Scotland.
Allen, Joan de Ris. (1990) *Living Buildings*. Camphill Architects, Scotland.
Ansell, Vera; Coates, Chris; Dawling, Pam; How, Jonathan; Morris, William and Wood, Andy (eds.) (1989) *Diggers and Dreamers*. Communes Network Publication, UK.
Arnold, Eberhard and Merton, Thomas. (1995) *Why we live in Community*. Plough Publishing House, USA.
Bailey, Keith; Bang, Jan and Matthews, Bob. Eds. (1986) *The Collective Experience*. Communes Network, UK
Bettelheim, Bruno. (1971) *The Children of the Dream*. Granada Publishing, UK.
Bock, Friedwart. Ed. (2004) *Builders of Camphill*. Floris Books, Scotland.
Brink, Margarete van den. (2004) *Transforming People and Organizations*. Temple Lodge, UK.
Budd, Christopher Houghton, and Steiner, Rudolf. (1996) *Rudolf Steiner Economist*. New Economy Publications, UK.
Christian, Diana Leafe. (2003) *Creating a Life Together*. New Society Publishers, Canada.
Coates, Chris. (2001) *Utopia Britannica*. Diggers and Dreamers Publications, UK.
Communities Directory, 2005 edition. Fellowship for Intentional Community, USA.
Critchfield, Richard. (1981) *Villages*. Anchor Books, USA.
Dawson, Jonathan. (2006*) Ecovillages — New Frontiers for Sustainability*. Green Books, UK.
Directory of Intentional Communities. (1991 edition) Fellowship for Intentional Community, USA.
Eno, Sarah and Treanor, Dave. (1982) *The Collective Housing Handbook*. Laurieston Hall Publications, Scotland.
Fike, Rupert. Editor. (1998) *Voices from The Farm*. Book Publishing Company, Tennessee, USA.
Foy, Nancy. (1980) *The Yin and Yang of Organizations*. William Morrow and Company, Inc., New York.

Gaskin, Ina May. (1977) *Spiritual Midwifery*. The Book Publishing Company, USA.
Gelb, Saadia. (1994) Almost *one hundred years of togetherness*. Kfar Blum Guest House, Israel.
Glasl, Friedrich. (1994) *The Enterprise of the Future*. Hawthorn Press, UK.
Gorman, Clem. (1975) *People Together*. Paladin, UK.
Groh, Trauger and McFadden, Steven. (1997) *Farms of Tomorrow Revisited*. Biodynamic Farming and Gardening Association. USA.
Handy, Charles. (1991) *Gods of Management*. Arrow Books, UK.
Hoehnle, Peter. (2003) *The Amana People*. Penfield Books, USA.
Holloway, Mark. (1966) *Heavens on Earth*. Dover Publications, USA.
Jackson, Hildur. Ed. (1999) *Creating Harmony*. Gaia Trust/ Permanent Publications. UK and Denmark.
Jacobsen, Rolf. (2001) *Tun, bygninger og økologi*. (Courtyards, buildings and ecology) Landbruksforlaget, Norway.
Kant, Joanita. (1990) *The Hutterite Community Cookbook*. Good Books, USA.
Leon, Dan. (1964) *The Kibbutz*. Israel Horizons, World Hashomer Hatzair, Israel.
Luxford, Michael and Jane. (2003) *A Sense for Community*. Directions for Change, UK.
Mindell, Arnold. (1995) *Sitting in the fire*. Lao Tse Press, USA.
Pitzer, Don. (1993) 'Developmental communalism.' Article in *Diggers and Dreamers*. Communes Network, UK.
Saunders, Nicholas. (1975) *Alternative England and Wales*. Nicholas Saunders, UK.
Schein, Edgar. (1999) *The Corporate Culture Survival Guide*. Jossey-Bass publishers, San Francisco, USA.
Schwarz, Walter and Dorothy. (1987) *Breaking Through*. Green Books, UK.
Steiner, Rudolf. (1992) *Ideas for a New Europe*. Rudolf Steiner Press, UK.
Steiner, Rudolf. (1974) *The Inner Aspect of the Social Question*. Rudolf Steiner Press, UK.
Van Diun, Veronika. (2002) *A Child in Community*. Upfront Publishing, UK.

Index

Ecovillages: A Practical Guide to Sustainable Communities

by Jan Martin Bang

Jan Martin Bang explores the background and history of the Ecovillages movement, which includes communities such as Kibbutz, Camphill and Permaculture. He then provides a comprehensive manual for planning, establishing and maintaining a sustainable community. Issues discussed include leadership and conflict management, house design, building techniques, farming and food production, water and sewage, energy sources and alternative economics. The final chapter brings it all together in a step-by-step guide. Includes over twenty 'Living Example' case-studies of communities from around the world.